Principles of Modelling and Rendering Using
3D Studio

Stuart Mealing, Brian Adams and Martin Woolner

Swets & Zeitlinger Publishers

Lisse, Abingdon, Exton (PA), Tokyo

intellect™

EXETER, ENGLAND

Paperback Edition First Published in 1998 by
Intellect Books, School of Art and Design, Earl Richards Road North, Exeter EX2 6AS

ISBN 1-871516-70-6
Copyright ©1998 Intellect Ltd.

A catalogue record for this book is available from the British Library

Cloth Edition First Published in 1998 by
Swets and Zeitlinger Publishers,
Hereweg 347B, 2161 CA LISSE, The Netherlands

ISBN 90 265 1523 5
Copyright ©1998 Swets and Zeitlinger Publishers.

Library of Congress Cataloging-in-Publication Data

Mealing, Stuart.
    Principles of modelling and rendering with 3D Studio / Stuart Mealing, Martin Woolner, Brian Adams.
      p.   cm.
    Includes bibliographical references and index.
    ISBN 9026515235 (hardback)
      1. Computer animation. 2. 3D studio. 3. Computer graphics.
  I. Woolner, Martin. II. Adams, Brian.
  TR897.7.M43 1998
  006.6'96--dc21
                                            98-4539
                                               CIP

Consulting Editor:	Masoud Yazdani
Copy Editor:	Lucy Kind
Cover Design:	Sam Robinson
Production:	Stuart Mealing
	Brian Adams

Printed and bound in Great Britain by Cromwell Press, Wiltshire

CONTENTS

Contents

CHAPTER 1

INTRODUCTION

There is a particular magic about creating and moving objects in three dimensional space of which one never tires, and an enthusiasm for the discipline which we hope to share with the reader. Computer modelling is an exciting area with which to become involved. Computers have not only revolutionised traditional modelling in many cases, but have allowed excursions into areas of complexity which have never before been possible, and have made these journeys accessible to users without the previously required specialist skills.

This book outlines computer modelling, rendering and animation and looks at their application using Autodesk 3D Studio, probably the most widely-used modelling program in the world. As with any book involving computing it is safe to assume that some hardware and software details will be out of date even before publication but much of our text deals with fundamentals that will remain unchanged. Even the sections introducing 3D Studio are designed to inform the user about general working practice and, as such, to supplement the manuals on particular versions of the software. Whilst it is not possible to write with total confidence about hardware and software that is not yet available, we can confidently assume that computers will grow faster, cheaper and more powerful and that these improvements will both broaden the range of people building computer models and increase the sophistication of the models they can build.

We have tried to produce a book which can be useful not only to the complete novice but also to the experienced modeller who is new to 3D Studio and to the experienced PC user who is new to modelling. As a result there will be chapters which some readers will feel qualified to pass by. We have also chosen to divide the book into two clearly separate parts, dealing firstly with generic theory of the discipline and then with the specifics of 3D Studio itself. In doing so we have allowed a little descriptive overlap between the two sections in order that each can make full sense on its own.

The book deals, in some depth, with the principles involved in modelling, rendering and animation as this theoretical knowledge helps to make sense of the practical issues of building a model on a computer. Apart from anything else, when you appreciate how much work the machine is having to do to produce a full-colour, high resolution, photo-realistic rendering of your detailed model of the universe, you might begin to forgive the delay in bringing it to screen! We also hope to show what is possible on the PC in different modelling categories – architecture, for example, being likely to make very different demands from those of animated TV advertisements. Having described the functionality and practical use of 3D Studio we have gone on to cover some of the new developments that have arrived for 3D Studio users with Studio MAX and Studio VIZ.

We three authors work together at the Exeter School of Art and Design in the University of Plymouth where we are members of the Centre for Visual Computing. The contributions to the book from each of us draw on our different skills – in Part One Stuart Mealing deals with theory whilst in Part Two Brian Adams and Martin Woolner, who use and teach 3D Studio daily, deal with the specifics of the program and illustrate its use through case studies. Each of us firmly believes that the recent advent of available 3D computing will enrich and extend the world of visual computing and will impact on many areas of our lives.

CHAPTER 2

THE WORLD OF 3-D

It would be easy to start writing about the 'How?' and the 'Why?' of computer modelling without stopping to mention the 'What?'. Consider briefly, therefore, the fundamental nature of the processes we are to look at and the new tools they provide.

2.1 WHAT IS MODELLING ?

Modelling might seem like an exciting thing to do, but what exactly is it and why is it so different from merely drawing?

Since you are already reading this book, you are likely to be sufficiently interested in the subject to have considered the above question, but many people have an initial difficulty in making the conceptual jump from two dimensions to three and this, therefore, seems a good place to start. The world around us is three-dimensional, having height, width and depth. Although we can talk about a picture being two-dimensional, in the real world the drawn marks are wedded to a surface which has a measurable thickness and a locatable position in three-dimensional space. It also has an existence in time, which is often termed the fourth dimension, and this temporal dimension is entered if we need to animate something. In order to design for this world, to imitate parts of it, or to create some other imaginary world, we need to be able to work in all three spatial dimensions. This is unavoidable when working with real materials – sawing wood, building with bricks or moulding clay for instance, as we manipulate them with our hands, but if we need to develop our ideas using a two-dimensional medium, such as in a drawing or on a computer screen, then we need to adopt an appropriate visual language.

This might appear intuitively obvious, but we see the problem from a Western, post-renaissance viewpoint where perspective is understood and photographs are familiar. It does, however, require a degree of intellectual sophistication to interpret flat images which describe the three-dimensional world.

2.2 NAVIGATING A 3-D COMPUTER WORLD

How is it possible to navigate a three-dimensional environment when we are looking at a two-dimensional monitor screen? Perspective offers a lot of assistance and the overlapping of solid objects indicates their relative spatial positions although real world depth clues are often much more subtle. Spatial depth can be indicated by the diminution of scale of texture in a scene, though aerial perspective is more difficult to evoke. This change in the apparent colour and clarity of distant objects, due to atmospheric influences, can be simulated only at the expense of much computing time. We will look at the way objects can be represented on the screen in the section on viewing models.

Many 3-D modelling packages offer three or four different views of the current scene on the screen at once in separate windows. Top view (plan), front view (front elevation) and side view (side elevation) for example, offer three co-ordinated diagrams which allow spatial positions to be evaluated accurately though not intuitively. Each of these two-dimensional views is parallel to the plane of the monitor screen and the cursor can, therefore, travel around it to access the contents in an understandable fashion. The choice of views can usually be set to determine whether the view is of top or bottom, front or back, left or right, and additionally a 'camera' view gives a perspectival view (as if looking through a camera viewfinder) from a defined point in space which corresponds

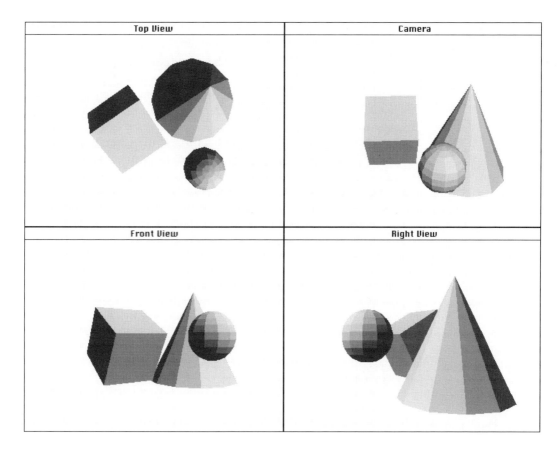

Top View	Camera
Front View	Right View

to the viewer's imagined position. These windows can be redrawn as close to simultaneously as the application will allow, though inevitable delays can make interactive navigation tedious and it is common to use abbreviated versions of the object(s), when possible, for the sake of speed. Some applications offer a single, perspectival view of the current scene which immediately raises the question of how movement in the extra dimension can be made clear.

The means of achieving this varies according to the application, but one solution is for the mouse alone to control movement in one plane (i.e. in two dimensions) and for the mouse, when used with a prescribed key held down, to allow movement through a plane at

Typical four window modeller interface

5

right angles to the first (i.e. the third dimension). This is only one of a number of solutions in use, and it tends to take a period of familiarisation before any method becomes comfortable to use. The ability to seem to enter into the scene itself would be very helpful and Virtual Reality, which is mentioned later but is currently at a crude state of development, will offer that possibility at some time in the near future.

An additional complication with a three-dimensional environment is that you can only see all of the scene if you are outside it. Once you have moved your viewpoint into the scene you may find that objects lie behind you, and if you have stayed outside the scene but magnified your view by 'zooming in', your reduced field of view may have cut off your sight of all objects. Most interface tools are not expressive enough to deal with these problems in an intuitive way and since monitors don't have wing mirrors it becomes a little like negotiating a room you are in whilst looking through a cardboard tube. Indeed it has been instructive, in the course of researching a range of 3-D modellers, to discover how many different paradigms there are for satisfactorily describing and navigating a three-dimensional world on a computer screen.

2.3 WHAT IS RENDERING ?

Our first impression, or mental picture, of computer modelling might well be of transparent, line-drawn, geometrical objects in an endless grey vacuum, a world of mathematical scenes and algebraic surfaces. This description of the generation of computer models suggests a heavily diagrammatic representation of any real world, and in order to create a realistic image of the world our objects need more treatment than just having their hidden surfaces removed.

Objects in the real world are illuminated by light of different colours and qualities, coming from a range of sources and directions - they cast shadows, they can reflect the world around them, they can exhibit different degrees of transparency, they have different surface qualities and we will need to be able to simulate these varied, textured and patterned surfaces. The interaction of all these qualities can give us a rich understanding about models and the scene they inhabit.

It is now possible to create computer-generated images (CGI) of stunning, photographic realism. Looking at such images, however, one rapidly becomes aware that this super-realism is often achieved within a limited domain where clean geometry and pure light prevail; probably with man-made objects in an artificially lit interior. The search is on for ways of dealing with less amenable subjects and, as always, the search can be as much fun as the solution. It is my hope that readers of this book will contribute to that solution. A way of giving 'difficult' subjects a similarly life-like treatment will be found, and it is interesting to speculate as to whether the modelling paradigms available to us now will be adequate for the task, or whether a completely new vision will be required.

Computer modelling, however, is not just about mimicking the world. Indeed it is often our choice not to observe rules which are absolute in real life. For example, in our computer-generated world we can decide whether an object will cast a shadow and we can create scenes where some objects cast shadows and others do not, or where some light sources cause shadows to be cast and others do not. We can decide what will be seen in reflective surfaces and can apply different rendering principles to different parts of the scene.

Rendering is, therefore, about making visible the objects and their given surface attributes under the chosen illumination and is made possible by the application of mathematical rules which determine how light will interact with the scene. A number of rendering algorithms exist, with varying levels of sophistication, and operating at different levels of computational expense.

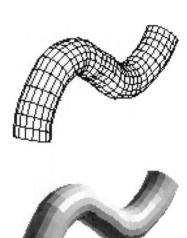

A model displayed in wireframe form (top), in quick shaded form (centre) and in smooth shaded form (bottom)

2.4 THE ROLE OF THE DESIGNER

The medium is properly used when it either extends the range of things the designer can do visually or makes easier, quicker or cheaper an existing part of the design process. Whether the presentation and manipulation of a 3-D logo is poetic or crass depends on the skill of the designer and the sensitivity of whoever commissions it; it is not a property of the medium itself. It is, however, often the case that a medium stimulates ideas and visions to grow in a particular direction. If the designer is constrained by the hardware and software available, with a machine designed for typographical aerobatics and with a 'chrome' rendering option, then the results may be predictable and the existence of those features on his machine will have been partly the result of market forces. The good designer will always be breaking new ground and consequently pushing hardware and software to its limits, but he will rarely be in a position to write software to push beyond the current limits of his application. There is, therefore, at graphics' leading edge, a growing liaison between designers, programmers and computer scientists.

CHAPTER 3

APPLICATIONS OF 3-D

There are unlimited applications for computer modelling and it is not practical to try to list them all. It is, however, worth isolating a few disciplines that make use of the medium and then to use them as examples to consider how their requirements can differ from one another. At this stage we are looking at the broadest range of uses of modelling, and in some specialised cases this discussion takes us beyond off-the-shelf applications currently available for the PC. It is likely, nevertheless, that 3D Studio running on a good PC will allow most things to be created.

The increasing ability to produce 3D computer animation at an acceptable cost and speed, and to employ it on a wide range of machines, is opening up many new opportunities for that particular branch of the medium. Almost everyone in the western world is being regularly exposed to the medium through commercial and entertainment uses on television, with dreaded 'flying logos' swooping past the eyes at frequent intervals. This increased exposure leads, of course, to increased familiarity and then, as the medium is accepted along side more traditional ones, to increased demand. Things in the real world are constantly moving, and the ability to mimic or simulate that quality breathes life into the inanimacy of the frozen image. A single image can capture 'the decisive moment', which might have become lost during a sequence, but many situations demand greater truth to turbulent reality.

3.1 ARCHITECTURE

Probably the most obvious single use of computer modelling is in architecture. Towns and buildings are usually straightforward to model on a computer, and this ability is increasingly utilised by architects, not only to experiment with different structures but also to demonstrate their choices to the client before large amounts of money are spent. Having modelled a proposal within its local environment, it is then possible to move around the model, viewing the building from any position, viewing the surroundings from within the building and assessing the total physical relationship of the building to its surroundings. The shadows cast by the building, by its neighbours and by trees on site can all be anticipated with far greater ease than previously possible, and on a sophisticated model it might be possible to simulate airflow around and through the new site.

It is equally possible to travel through the building to preview the internal appearance and layout, to try different permutations of lighting, different decors, changes of ceiling height and position of windows, for instance. The same database from which the model is constructed might be accessed by an expert system to calculate percentage area of windows, heat loss under different conditions, and conformity to changing building regulations. Animation offers the possibility of anticipating traffic flow problems by watching them develop on screen, or of seeing the shadow from the new office tower creep round to engulf the nearby housing estate, and it is clear that this will become a major planning tool. An architect designing a child care centre has already been able to 'test out' his design by moving around it 'as a child', in this case exploiting technology which enabled him actually to move like a child as well as see from a child's viewpoint.

In a simpler case, the remodelling of a foyer or a domestic kitchen can be previewed much more clearly by a client who has no experience of reading plans, than could be traditionally, where those plans were likely to be supplemented only by an artist's impression or single perspective view. A floor plan can be entered into the computer and 'extruded' to make a basic 3-D model in a few moments, after which you can 'move around' anywhere inside or outside that space. This offers the improved efficiency of making and viewing changes in company with the client and others in the design team. One reservation about computer modelling, however, is that it offers an immediate believability and finality at the sacrifice of a rough sketch's openess. There is also a superficial credibility about a clean computer model which might disguise flaws and design weaknesses from the layman. The rough sketch, however, is somehow pregnant with possibilities which the computer model has tidied out of the way, and it can be most fruitful when the two techniques can co-exist.

It may also be desirable to avoid the time and expense of building a traditional model, particularly at the stage of competing for a job, by substituting a computer model. This can be a major economy if the plans are already on a computer since that same data can be used to generate the computer model.

3.2 3-D DESIGN

The 3-D designer may be creating anything from a high speed train to a hairpin, but is increasingly likely to be doing it on a computer. This likelihood has increased since the introduction of much more intuitive modellers, which allow looser, more 'sketchy' ideas to be translated into three-dimensional objects. A problem still exists about matching the flexibility and openness of a thumb-nail sketch on the back of an envelope with a formal 3-D modeller, but the application of artificial intelligence in pen-driven devices could offer a clue to bringing the two together. It might, however, prove to be inadvisable to try and have the computer guessing at the meaning of a scribbled drawing, since it can be the very ambiguity of the sketch that provokes fresh ideas to develop.

As a design matures, it becomes increasingly important that it can be understood unambiguously, and a standard modeller can bestow complete mathematical accuracy on the three-dimensional design. When the process moves from design to manufacture the computer model can be passed on down the line to be used in the design of production tools and can possibly be used for ergonomic or stress testing.

A packaging designer will need to be able to wrap flat images around the product and its container and will probably then need to be able to insert the product either into a simulated environment or into a photograph of one. These requirements bring the need to integrate 2-D and 3-D material in the same environment. Finally, the model might go on to be used in an animated TV advertisement for the product.

A further development might be for the manufacture of the object to be controlled by the same program that created the model. In CAM (Computer Aided Manufacture) the computer is connected to the machine (such as a lathe) which produces the component and the data describing the object ,which could have come from the CAD (Computer Aided Design) half of the same package, determines its operation. A number of production methods exist for getting from the data to an object; the object itself perhaps being a prototype model which will provide fresh information that is in turn used to modify the original design. This is the world of CAD/CAM in which rapid-prototyping is becoming increasingly available as the 3-D equivalent of a laser printer, though desk-top versions are currently more than a hundred times the cost.

3.3 ENGINEERING

The designer of a high speed train will be one specialist working in a team with other specialists such as engineers. In the construction of a bridge or oil refinery, however, the design function may rest with the engineer alone. The engineer might also require a model which gives access to specialised information, in order to assess component stress for example, and might use techniques such as finite element methods to that end. Although a very specialised field, it is interesting to note how the model can be constructed to provide specific and measurable feedback to its creator.

It is also worth considering to what extent a discipline such as engineering might be (and has been) impacted by computer modelling. In aeronautics, for example, engineers can now study the effects of stresses and strains on the airframe by simulating meteorological extremes, G-forces etc., and subsequently check modifications against the same conditions. This leads to an understanding of the operational limits and to the definition of the aircraft's flight envelope.

RFT (Right First Time) engineering is gaining converts, particularly in industries such as aerospace which have high development costs. Firstly, initial specifications are run through a computer to identify early errors and thus make cost savings. Secondly, design and testing procedures are enacted on screen when possible, in order to minimise the need to make real prototypes and to shorten the time it takes to reach a final product. The latter stage is largely dependent on 3-D computer modelling.

3.4 SIMULATION

Visualisation can lead to revelation. Several hundred years ago, overlaying the location of deaths from cholera, on a map of available water pumps, traced the cause of a London epidemic. Held separately the two pieces of information yield nothing, but the importance of the knowledge gained from combining the two, explains the search for increasingly sophisticated methods with which to draw conclusions from separate pieces of data. The need for this development has been accelerated by the quantity of data which computer technology can generate, and the impossibility of making useful judgments about it. It had been pointed out that storage capacity increases were not keeping up with those of computational speed, leading to the conclusion that a researcher could compute more than could be stored and could store more than could be comprehended.

A graph showing acceleration, for example, plots speed against time, and has a clarity and immediacy which is lacking in the raw data from which it is constructed. A two-dimensional graph shows the relationship between factors whose proportions are indicated on two axes (i.e. plotting X against Y, plotting house prices against year). A three-dimensional graph extends the factors that can be compared by adding a third axis (i.e. plotting X against Y and then extending into Z, plotting house prices against year in different regions). The information from several 2-D graphs can thus be condensed into one 3-D graph with a potential increase in clarity. It is necessary to be careful about the scale of axes to preserve accuracy, and to find an appropriate presentational form to prevent 3-D information being obscured. The 3-D contour map which can be created in a 3-D graph, can have its surface overlaid with a further layer of data, effectively creating a 4-D image. It is also possible to plot diagrammatic information over a 'realistic' 3-D form, such as overlapping an operating temperature map for a disc brake over a 3-D model of the disc.

An American firm specialises in creating animated computer simulations for use in lawsuits. It recreates car crashes which have involved the litigants, incorporating parameters based on those present in the actual accident in order that the incident can be studied in court. This is in accordance with one definition of simulation: the reproduction of the conditions of (a situation etc.), as in carrying out an experiment. It is more problematic as a piece of legal evidence if an alternative definition of simulate is tried: to make a pretence of, to feign. A simulation must embody truth about the situation it seeks to reproduce but at the same time need not pretend to be that actual situation. Whilst recognising that we are looking at organisations of pixels denoting two automobiles on a flat screen, we can derive useful information about what two real vehicles would do in a given situation, providing the representations have been programmed to make accurate responses in terms of the masses, forces and frictions involved in real life. Simulations seek to model reality with different levels of fidelity. This is modelling 'plus', since the model has not only spatial dimensions but also attributes that affect its performance in the temporal dimension, such as mass and centre of gravity. These features only come to life in an animation.

As well as being able to recreate an incident from the past, it is practical, and more usual, to want to create a simulation of a theoretical event. What would happen if one of the cars had been travelling twice as fast? At what point would a bearing fracture if it were put under an increasing load? By providing the right forces to a model which 'knows' how to respond, we can watch the event unfold before us, then vary the parameters and observe the changes. This also allows us, in the right circumstances, to build and animate a scene by describing the physical rules which will apply, rather than having to control manually every element.

A special case of simulation is the flight simulator. A 'top-of-the-range' flight simulator will model the experience of flying an aircraft with such accuracy that airsickness can be a genuine problem. At the cost of several millions of pounds, the pilot can sit in the aircraft of his choice, confronted by an authentic cockpit display, with a full set of 'working' controls, a realistic view of his chosen airport visible through the windscreen, appropriate engine noises, and can 'fly' the plane in any chosen conditions, with the correct flight characteristics. Hydraulic rams under his 'cockpit' tilt and rock him just as a real aircraft would do, and the combination of physical and visual stimuli is so convincing that it is necessary to concentrate very hard to question the reality of the flight experience. In some military simulators, the addition of snug hydraulic suits through which pressure can be increased on the body, and seat belts which can exert sudden tension on the pilot, also allow the stresses of acceleration and increased G-forces to be reproduced.

Even the relatively crude visual display of a flight simulator on a home micro is considered, by qualified pilots, to have a useful level of realism, and a high-end workstation even more so. It is now permissible for trainee pilots to log hours towards their licence on inexpensive software which is available for the home market as games. Flight simulators, of course, are more than just sophisticated fairground rides. They save aircraft, lives and money by allowing for efficient ground training, where landings and take-offs from obscure airports can be practised repeatedly, responses to in-flight emergencies rehearsed and pilots 'converted' to new types of aircraft. Military pilots can practise bombing runs, in-flight refuelling and landings on aircraft carriers without risk of dangerous and expensive mistakes. These principles can also be applied to other types of vehicle and equipment – locomotive cabs, oil tanker bridges and anti-aircraft guns can all be simulated using similar techniques, and in each case the operator's world is produced by computer modelling.

It is even more difficult to rehearse something which will happen half way across the solar system, and so space research makes heavy use of simulation and visualisation. The resulting material is also important in the fight for project finance, and it has been suggested that the fine computer-animated previews of the

Voyager spaceprobe played a big part in winning funding for the mission. The quality of movement of objects in space - that smooth, slow, cleanly defined pace - seems well matched to the sort of motion which computer animation produces most easily. Its silky accuracy often looks odd when applied to earthbound activity, but outside the earth's atmosphere everything appears to move like a flying logo. The particular clarity, and spatial depth of images from space, with its limited number of light sources, is well mimicked by the computer, and it is also convenient that most of the man-made objects, which are the subject of these animations, are constructed using the geometry which computers most readily generate.

The Voyager example does bring to light an interesting question about the ethics of changing things to make them more visible. If a 20-hour flypast of Jupiter is condensed into 3 minutes, how true to the real event can the simulation be said to be? Similarly, it is often desirable to change the contrast ratio of an image to facilitate its reproduction in a newspaper, or to change the colour range to suit television reproduction, but this could be seen as tampering with evidence on which scientific judgement is based. When images are returned from distant places in the universe, the colours used in their reproduction are likely to be altered in order to make certain features more visible, and any notion of a 'true' record must be balanced accordingly.

It is particularly important for astronauts to have access to simulators of the vehicles and conditions which space will present, since the moon is not a good place to make a first attempt at flying a lunar module! Specialised variations on flight simulators provide that opportunity. Space scientists can also rehearse proposed trajectories without the risk of losing a valuable payload, and data from unmanned space missions can be used to generate authentic looking flights over the surface of distant planets prior to manned landings. The construction of space stations can be rehearsed, amended, demonstrated and practiced in an environment where gravity can be switched on and off at will.

Our interest is centred on the scene presented in the visual display, and a number of clever shortcuts may be implemented in order to be able to move realistically through a scene in real-time. Dusk and night simulations require less detail to feel realistic, and point light sources alone (which are easier to manipulate than polygons) may provide much of the visual information about an airfield at night. Instead of 'building' a city from polygons it might be possible to 'stick' pictures of the city onto highly simplified shapes. Similarly it is sometimes appropriate to produce authentic looking clouds by 'sticking' cloud pictures onto simple blocks, and shadows will often be acceptable if they are exist as silhouettes on an idealised ground plane, though failing to adapt to contours and obstructions.

3.5 MEDICAL

The ability to extract data from scans takenof patients and construct from it three-dimensional computer models is proving an important new diagnostic tool in medicine. Previous technology only presented two-dimensional pictures of internal structures, and it was necessary to resort to surgery in order to confront organs in three dimensions. This new method makes it possible to build skulls, vertebra, hearts and brains in the computer and then to manipulate them on screen. Volume visualisation (described in Chapter 4) permits a three-dimensional model of a body to be peeled back in layers to reveal the relevant organ. Any ambiguities about the exact form can be further removed by animating the part, though it is likely that this will not (currently) be in real time if using a highly detailed model.

In all cases of medical imaging, and indeed any specialised area, it is important to recognise that the computer operator must be working with someone who knows what is being looked for and what needs to be seen. Whilst you and I can look at a computer model of a group of articulating vertebrae and be impressed with the clarity with which their movements are shown, the animation is medically useless if it does not reveal what the doctors need to see. It is the person with medical skills who must decide what is needed and the job of either the system or the operator to manipulate the data to provide it. Increasingly friendly and intelligent systems make it likely that the doctor and the operator will be one and the same person, but at the current stage of development that is unlikely to be the case.

The reconstruction of shattered bones or rebuilding of a

deformed skull involves a three-dimensional jigsaw that can be rehearsed on the computer model. Also of assistance, to plastic surgeons in particular, is the ability to experiment with manipulations at a model stage, for instance checking on a model that bone removed from the foot can be used to build a part of a jaw. General operation simulators are being developed which will permit doctors to practise surgery in simulated three-dimensional reality. This idea will soon extend to operations being carried out by doctors hundreds of miles from the patient and will be a serious part of future space flight scenarios.

3.6 TV GRAPHICS

Although there is nothing unique about the computer modelling and rendering techniques used in TV graphics to distinguish them from those used in most other areas, they are, however, the most public manifestation of the art. TV graphics have been taken on board so readily by producers and designers wanting their programme introduction or promotion to have more punch than its rivals, that it has almost become the de facto standard. They have also become heavily used by television advertisers, notable for the size of budget they can sometimes make available. As a spin-off, this has unfortunately brought to millions of people, in the privacy of their homes, some of the most vulgar and needlessly expensive images of the century. The best examples of the genre have, however, become minor classics which enlighten and contribute to the discipline of graphic design.

Quantel has been the name associated with computer paint systems for a number of years and is still a yardstick against which others are measured. It may or may not be the best, but it has been clearly identified with the revolution in TV graphics, and helped moved designers to a more central role in production. When it first became available designers at the BBC found that it so considerably speeded up their jobs that they were prepared to forsake their families and work during the night if that was when the machine was free. Its relevance to animation was less marked until the Harry system was coupled to it. Harry is a digital editing suite which gives enormous flexibility in the manipulation of images from a range of sources, including live action video, without the generation loss which inhibits normal video work. It allows an animated sequence to be worked on (either frame by frame or in its entirety), added to and processed indefinitely without loss of quality, and with intuitive ease. Since these two pieces of kit started to transform TV design practice, the performance of PCs has increased to the point where they are used both alongside, and sometimes instead of, the more expensive hardware. Expense is, of course, a major factor, as a number of PCs running 3D Studio can be set up for the cost of one heavyweight system and offer access to many designers.

It is hard to generalise about the use of computer modelling on television as its function and form will vary according to the context. Modelling technology has effectively provided an additional step in the evolution of video graphics and there are several areas where it is currently popular. A seminal example from UK television is the Robinson Lambie-Nairn ident (a logo, motif or graphic sequence to identify a station) for Channel 4 which was created in 1982 and is still running more than ten years later. Station idents, programme title sequences, information graphics and advertisements all make heavy use of the medium and it is almost universal, at the moment, for news programmes to employ computers in the production of their introductions. News programmes are something of a flagship for the stations and are an important way to establish their house style. The graphics may need to evoke qualities of honesty, seriousness, topicality and grittiness, define the relevant locality, reinforce the station's image and be accompanied by a matching soundtrack. The images used are usually iconic (the globe, the parliament building), the typography prominent, the animation smooth and pacey, and the overall feel often symbolic (reaching out across the airways, flying to the nation's pulse).

Television weather forecasts usually employ a range of computer animated material, in addition to their 'intro', and are interesting in that the weather charts themselves need to be remade, perhaps several times each day. A system is therefore required which will allow the rapid production of fresh images from meteorological information. This might be the animation of digitised satellite photographs, moving isobars or 'raining' clouds and 'shining' sun icons. This means that information must be received at regular intervals from a meteorological source, and that there must be a quick method of getting from production of the charts to the point of broadcast. One method allows the charts to be compiled on PCs using a customised library of icons, which then automatically controls a Quantel paintbox in down-time to produce top quality graphics.

Strings of 3-D letter forms can be seen to lend themselves to geometric manipulation in space and are able to retain a high degree of legibility throughout major transformations. Such manipulations are well within the ability of PC hardware and software and the PC is increasingly finding itself a place in the production environment. It often gives way, at the moment, to more specialist machines for a number of tasks, including heavy rendering operations, but is considered to be very cost effective. The video handling capacity of the newer PCs is also attracting great interest from TV production departments.

3.7 SPECIAL EFFECTS

Special effects (FX) can be a special case for TV and film graphics. As the credits roll on many feature length films today, reference will be seen to computer special effects. The ability to generate impossible visions 'realistically' is all in a day's work for the computer and has come to be widely exploited. Until recently the classic examples have been in space films, where computer modelled spacecraft, planets, meteorite showers and the like can be created and choreographed with some ease, often intercut or merged with live or model shot material. Increasingly, however, computer generated effects have become almost standard and films like *Jurassic Park* rely heavily on them throughout. One advantage of computer generated sets, as opposed to hand-built models, is that they can be destroyed as often as you like and then restored at the touch of a button. This has to be set against the additional time currently taken to construct and render a complex computer model, though improving hardware and techniques will soon give the computer method a clear edge.

It had been estimated at Industrial Light & Magic, an American company renowned for special effects production, that in 1989 only about 2% of their effects currently used computers and that whilst that percentage would increase, they were not expected to take over entirely from model makers. Ten years on from that prediction I suspect that the percentage has multiplied by at least ten, as showcased by the digital crowds in the recent film *Titanic*. Top model makers have honed their skills over a number of years and one of their stocks-in-trade is dirt and the ageing of models, which often seems alien to computer graphics programmers, and is not always easily implemented when required. It is also difficult, at the moment, for computer models to match the subtlety of lighting that exists on a real set, and the primary requirement of special effects is that they <u>must</u> match the look of the rest of the film. A major advantage of computer graphics and animation, however, is that the 'virtual' camera and lights have zero dimensions. There is nowhere that the computer camera cannot go, no gap is too narrow for its passage and it can pass through walls to order. Similarly, scenes can be illuminated without the physical presence of real lights to contend with, so there are no cables to hide, nothing to keep out of shot, and no problems with heat or power.

In *The Abyss* a remarkable special effect from Industrial Light & Magic models was a pool of water growing an arm-like tentacle which retained all its clear, reflective and transparent properties while it extended, moved towards actors, transformed its end into a face, and is touched by an actress. Its smooth, gently rippling motion made it visually idenyical to water and yet able to do things wholly impossible for water. The brilliant sequence took six people, with the assistance of part-timers, six to eight months to produce 75 seconds of film (close to one second of animation per person per month). It also took nearly five hours to render each frame, with a number of steps to ensure that fog, shading, reflection, refraction and highlights were all correctly shown. By coincidence, the research team at London's Electric Image was developing a similar effect at the same time, which serves to suggest that the leading edge of the discipline is internationally spread.

Transformations now commonly use digital technology to advantage and are quite common in fantasy films where a frog might metamorphose into a prince, for instance, or an ice-cream. Several fairly recent films, such as the *Terminator* series, raise the technique to new heights, and the process of 'morphing' from one 3-D model to another is not difficult in principle.

3.8 ART

Until recently much computer 'art' has been poor, often because it has not really been art at all but merely the visual product of computer scientists' experiments. There are, however, signs that the medium is improving and the computer is rapidly finding a place as a tool for artists. It has had its teething troubles (in the same way that photography did) but is starting to establish its own unique identity. William Latham has created sculptures on a computer which could not exist in real life, and the obvious way to view an imaginary sculpture is to move round it in a 3-D model. He uses constructional solid geometry and texture mapping (described in Chapters 4 and 5), to create delicate, magical structures sometimes resembling hallucinogenic seashells, his software being custom written with a colleague. These forms are variously presented as photographs, on computer screens or in animations where the viewer is 'flown' through the intricate, coloured tunnels of the sculpture without the inhibitions of gravity or reality. The ability to waive the laws of gravity and to create 'impossible' objects in 'impossible' environments is potentially attractive to the artist, and the new-found availability of the means of such creation suggests fast development in the area.

The mathematical basis for some forms of art (remember 'op art'?) leave it open to obvious development by computer. This readily applies to work in 2-D and 3-D, where there has been a consistent interest for a number of decades, but it can also be extended into the fourth dimension with animation.

3.9 FILM ANIMATION

There is a growing use of computer animation, particularly of 3-D animation, and this must necessarily be preceded by building the 3-D models involved. Indeed, such animation can be broken down into three stages – modelling, choreography and rendering. Although computers are increasingly used in 2-D (and 2.5-D) animation, it is with 3-D that their presence is most significant, since it is here that they permit things to be created which could not be done any other way. The uses of 3-D computer animation are many and growing, stretching from Scientific Visualisation through education to entertainment.

One of the great exponents in the field of entertainment is John Lasseter, who is an ex-Disney animator working with a team at Pixar in California, well known for the computer animations *Luxo Jnr.*, *Red's Dream*, *knicknack*, and the Oscar winning *Tin Toy*. My favourite of these is *Luxo Jnr.*, a miniature masterpiece in which the medium has become completely invisible and we enjoy the animation for itself. The stars are two angle poise (Luxo) lamps, mother and child, who act out a scene (in which the youngster plays with a ball watched by his parent) with a level of characterisation that is close to human. It is a classic example of the technology being handmaiden to the art, though in this case the technology has been developed to a very high level of sensitivity. Telling details include the understated set and palette (computer graphics too often has all the colour knobs set to maximum), the pinpoint accuracy of the few sound effects, and the proportioning of the child lamp. Instead of being a small version of the parent, it is proportioned in the same relationship of human child to adult: small light shade but same size bulb, shorter support rods and springs but with the same diameter. In this film the angle poise lamps lend themselves readily to computer modelling, being made up of geometrical shapes, and it is interesting to compare them to the human baby in *Tin Toy* where it is apparent that computers are far less willing to model a chubby, flexible child than spheres and cubes. Most recently the Pixar team has produced (to great acclaim both inside and outside the industry) *Toy Story*, the first, full-length, wholly computer-generated film.

3.10 ARCHAEOLOGY

It would be possible to pick on almost any discipline area and find applications within it for computer animation. This chapter, therefore, selects just a few. Since the examples given tend to be the more obvious ones, I include mention of a perhaps less-expected example of the use of the medium in the field of archaeology. An archaeological excavation involves the investigation of a 3-D space over a period of time, and the acquisition of large amounts of data. Computers have already proved their use in the management of the data that accrues, but the vital recording of continuing changes to the site, and the locating of finds, suggests a 3-D model able to reflect those sequential changes.

Paul Reilly has described a simulated excavation site named Grafland, of which he built a three-dimensional computer model showing soil layers with various features (such as pits and post holes) cut into them, which constitutes a record of the data inevitably destroyed in the course of excavation. An animation shows a green 'field' falling away to leave a block of ground which represents the excavation volume. This volume is manipulated to show various features – the major layers, sections through pits and post holes, buried items, etc. Individual features can be isolated and observed, a hypothetical artefact assemblage can be shown in situ, and layers can be removed in sequence or added in reverse sequence. The whole piece provides a graphic record of the site, and changes to it, which traditional methods would find hard to match.

Computers are also being increasingly used to construct models of buildings, and such like, from the parts revealed by excavations. It is much easier to hold components in the spatial relationships in which they are found in the gravity-less computer model, than in a real world model, and to subsequently manipulate them and, perhaps, change the model's scale. Much more complete structures, such as the Roman Baths at Bath, can be explained and explored with animated computer models, and are becoming a familiar educational resource at such sites. The reconstruction of artefacts from a complex jigsaw of pieces has also been facilitated by computers, the necessary spatial manipulations being made possible by modelling systems.

3.11 EDUCATION

The use of video material in education has grown with the technology, and it is a natural development that computer modelling and animation should become two of the production tools. The increase in specifically educational programs shown on television, such as the Open University in the UK, has created a market which can utilise both high-end and low-end animation. Sometimes the presentation can be simply business graphics, with bar charts and such like, but in a learning situation these basic visualisation techniques can be most valuable. At other times more sophisticated techniques may be appropriate, and whilst the educational budget is rarely high, if production times are less rushed then economies can be made. The product can also be expected to stay on the market for a number of years and benefit a large number of users.

A particularly inspired set of videotapes called *Project Mathematics!* has been produced by Jim Blinn (long-time computer graphics guru and past simulator of the Pioneer and Voyager missions) to teach high-school mathematics, with funding from several sources. It is, perhaps, easy to imagine how the mathematics underlying all of computer graphics could be readily employed in the service of explaining that same mathematics. How immediate the relationship between a viewing transform (which converts data about 3-D space in order to display it on a 2-D screen) and an animated demonstration of aspects of trigonometry. Once again, however, it is the coming together of mathematical and visual skills which proves so productive.

In common with other fields, educationalists are very interested in multimedia presentation, where sound, live video, still images, animation and text can all come together. The laser dish is the medium which has precipitated development in this area, though it might be overtaken by other digital media. Also the increased memory of the latest PCs, together with greatly improved data compression techniques, has lead to multimedia in a single, intelligent box. A particular advantage of this technology is that it need not be linear, and is rarely designed to be so. It is not switched on and followed from beginning to end, but is used interactively, with the user determining the route, and speed, taken through the information. Each user, therefore, effectively constructs a personal course according to individual interests and pace of learning, although hopefully under qualified supervision.

3.12 GAMES

Animation is almost a prerequisite of computer games and 3D
Studio has found a particular niche here. Whether it is Pac-man
gobbling up opponents as he traverses a maze, space creatures
advancing to be destroyed in a 'shoot-em-up' game, or just chess
pieces moving themselves in response to your move, games abhor a
static screen display. Because the display is attempting to be
interactive on a simple home computer, the complexity of the
moving image has to be relatively simple, but games creators take
great pride in optimising routines and hacking corners to improve
their performance.

   The big brother of the home computer game is to be found in
amusement arcades, where more advanced graphics on more
sophisticated hardware lets you crash cars and kill aliens much more
spectacularly. Arcades also have a brash, noisy atmosphere and add
a social dimension which enhances the games for aficionados.
Arcade games can be exciting, involving and even addictive.
Dramatic perspective, colour and speed are typical features, but
some of the latest machines borrow heavily from state-of-the-art
simulators to condense the sensation of landing a jumbo-jet, or flying
a spitfire in battle, into a small cubicle at a cost of a single coin.   The
realism is eerie as you battle with the controls of an aircraft coming
into J. F. Kennedy airport in the corner of a pub in Soho, and is still
credible sitting in your living room at the keyboard of your own
PC. On a grander scale, the *Body Wars* ride at Walt Disney World
EPCOTT Center in Florida simulates a journey through the human
body for the audience of a small theatre mounted on a hydraulic
platform. The ride is not interactive, but consists of 2 minutes of
computer animation, generated at film resolution, matched by the
movement of the platform.

   Other non-games, which involve little user input, are more like
house pet substitutes. One involves little computer figures
inhabiting a cross-sectional house on the screen, living their lives,
albeit rather restrictedly, for the entertainment of the user, whilst
another has computer fish swimming on the screen. Although
these games might not be very meaningful, a number of scientists are
creating stimulus-response animations, in which cellular automata

respond according to rules governing their behaviour. The rules can involve response to environment, to hunger, to population density, etc., and the social orders achieved can be controlled by varying the rule parameters and can be studied in relationship to those of real creatures.

3.13 VIRTUAL REALITY

Virtual Reality (VR) is a much-hyped medium which allows the user to participate in a computer-generated world. This participatory experience can be either immersive or non-immersive. In immersive VR, the user has the apparent experience of being within, and able to move around in, a computer-generated scene; the current enabling technology being a head set with TV screens in front of each eye (allowing the user to look around the scene) and data gloves and suit (allowing interaction with, and feedback from, the scene itself). Non-immersive VR provides a window through which the user can view the scene and the means to change the view of the scene and to interact with it (in practical terms this is likely to mean looking at a VDU display).

Architects are already adding functionality to computer models of proposed buildings by turning them into virtual environments which can then be navigated by issuing 'forward/back/left/right/up/down' commands, often using the ubiquitous mouse. This allows the architects themselves, as well as their clients, to move through and view any part of the building (on a screen) from any position. In fully immersive VR the user could navigate the building as if it really existed, walking forwards, turning left, looking up, etc.; also being able to touch and move virtual furniture, or even elements of the building itself, perhaps using an interactive glove with feedback.

This could be considered as more of a philosophical than functional step forward from the existing ability to manipulate, and interact with, a model on screen, but there is great potential for the application of VR. For example, as a means of robotic control in hazardous environments, or a means of doctors practising operations without using real patients, or (as has already been suggested) as a means of those doctors performing operations from Earth on astronauts in space!

Whatever the scenario, the computer model needs to be built and rendered, and companies are already specialising in this task. Relatively inexpensive programs now exist for the Mac which enable the creation, and non-immersive navigation, of models. As with most areas of computer animation, there is a trade-off between the complexity of the model and the speed with which it can be redrawn which provides an obvious impediment to smooth movement through a detailed scene. This will rapidly be overcome as subsequent generations of hardware become faster and more powerful, and meanwhile models can be built on a PC for export to full VR systems running on the fastest hardware.

CHAPTER 4

FUNDAMENTALS: MODELLING

An understanding of the basics of computer graphics theory helps immensely when it comes to using any software package. It is perfectly possible to use a well designed application without knowledge of the theory underlying its operation, but the additional insight given by knowing how and why things happen as they do offers the user greater control. The manual for the application is likely to explain, for example, that certain modelling or rendering options are very time consuming, but knowing why that is so not only gives a greater over-all understanding of the package but makes it easier to plan more efficient ways of using the package.

This chapter aims to give a compact introduction to those areas relevant to our subject and does not concentrate exclusively on the features available to us in 3D Studio, and includes brief mention of some other areas that we might be drawn into. It assumes that the reader knows little or nothing about the subject and can be by-passed by those who have dealt with the subject before, although it should still offer a basic source of reference. There are many books dealing with computer graphics theory in great depth and some recommendations are listed in the bibliography.

4.1 DISPLAY

It is easy to forget that the monitor screen is only one of a number of different possible output devices (output is dealt with at greater length in Chapter 8). It displays a representation of what is going on inside the computer to the best of its ability but should not be taken as an exact record of what the machine has calculated. It is possible, for instance, that a sphere is stored with complete mathematical accuracy in the memory of the computer but in displaying it on screen the limitations of resolution and palette size give us only an approximation of that perfect sphere. That might prove satisfactory in use but the increments available to us when navigating the screen are determined by the size of the pixels and these are unlikely to allow accurate alignment of parts of a model. In this case, numerical entry from the keyboard might be used to achieve the necessary accuracy, which the screen would then display as best it could. (The word *pixel* comes from shortening *picture element* and labels the basic dot-like unit from which a screen display is made. Typically, each square inch of screen might comprise 72 rows each of 72 pixels).

Enlargement of a letterform detail clearly showing the pixel structure

4.2 COORDINATES

In order to move about the screen, and for us and the computer to keep track of where we are (probably indicated by where the cursor is), a coordinate system is used. In the same way that a map reference locates a point as being at the intersection of imaginary lines drawn from one side and from the bottom (or top) of the map, so the pixel column and row numbers address specific screen locations. The horizontal axis is the 'X' axis and the vertical axis is

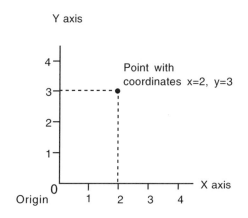

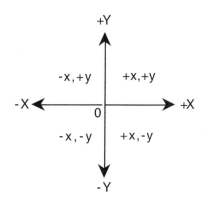

the 'Y' axis. On a display with 624 horizontal lines each of 832 pixels, and with the origin (the point defined by 'X=0,Y=0') sited at the bottom left of the screen, then the pixel addressed by the reference 'X=416,Y=312' would be one of the four pixels surrounding the point dead centre of the screen. These are the screen coordinates.

It is clearly necessary to distinguish between a position on the screen and the position of the model within its own world. If we zoom in on our view of the model then the screen co-ordinates referencing a point on the model (unless that point happens to coincide with the centre of the screen) will change, whilst the model

Left: 2-D cartesian coordinate system
Right: Positive and negative coordinate locations around the origin

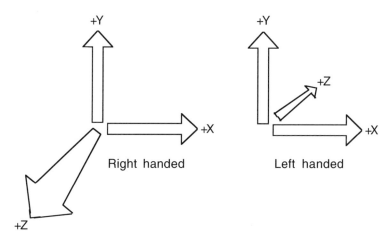

Right and left handed coordinate systems, showing the reversal of the positive Z axis

itself will not have moved within its own world. A second set of co-ordinates, therefore, describes the position of the model within its own world, and the computer converts those coordinates to screen coordinates when displaying the model. As the model is three-dimensional, an extra axis, the 'Z' axis, is needed to indicate depth. The model's world may be either infinite or limited along each axis. Since this world extends in all directions from its centre (where X=Y=Z=0), then each axis has a positive and negative direction. For various reasons, the labelling of the axes has differed between different disciplines but for our purposes the only inconsistency we might find is in the Z axis which is sometimes positive 'going away' from the viewer (known as left-handed) and occasionally positive 'coming towards' the viewer (known as right-handed).

4.2.1 RELATIVE AND ABSOLUTE

When modelling we are likely only to be presented with world co-ordinates, although the application might use other coordinate systems internally for its own purposes. These world coordinates are often presented as an XYZ readout in the margin of the current window (which updates as positional changes are made), and the co-ordinates of objects can often be accessed and modified through a specific object information box and often through a set key command. It can be convenient to establish a rough position by manoeuvring on the screen and then using numeric input from the keyboard for precise positioning.

Positions and movements can be described as either relative or absolute. An absolute position is a specific location in the object's world as described by the world coordinates. A relative position does not refer directly to the world coordinates but is located in terms of distance along X,Y and Z (or a combination of distance and angles) from a known point in object space. This is useful if we wish to move an object a known distance from its current position without wanting to specify the world co-ordinates of the new location.

4.2.2 CARTESIAN COORDINATES

The 16th century philosopher and mathematician René Descartes gave his name to the Cartesian coordinate system (which is the one referred to so far). This system locates a point by measuring along the X and Y axes in 2-D, and the X,Y and Z axes in 3-D from a given point of origin.

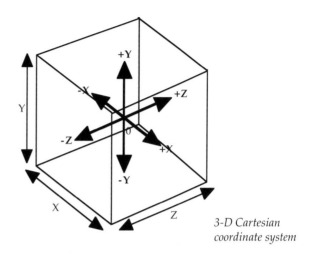

3-D Cartesian coordinate system

4.2.3 POLAR/SPHERICAL COORDINATES

An alternative to the Cartesian system is the polar coordinate system with which, in two dimensions, a position is located by its distance from the origin and the angle between the X+ axis and a line from the origin to the point. In three dimensions two angles plus distance from the origin are required, and the method is known as the spherical coordinate system. This can be an intuitive way of moving

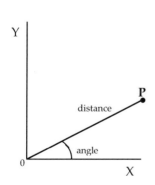

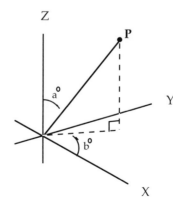

*Left: 2-D polar coordinate s
Right: 3-D spherical coordinates*

about a space as it often matches the way we think about changes of position. It is, for instance, a much more likely way to describe the flight of an aircraft to a location, but has not proved as clear a way of describing the location itself within currently screen-bound applications.

4.2.4 HOMOGENEOUS COORDINATES

Although not important to us as software users, homogeneous coordinates are mentioned here for the sake of completeness. They simplify the mathematics used to manipulate coordinates by the program, and require a three-dimensional point to be represented by a four number vector. If you look at any such calculations it can be confusing to see a point referenced by one number more than the expected three dimensions (e.g. the point x,y,z is represented by [x y z 1] to allow matrix multiplication).

4.3 RASTER

The image is formed on a normal display screen by a raster, which is a set of horizontal raster lines scanned along each successive row of pixels. A raster image is one which describes its subject in terms of the pixel intensities across its surface. It is, therefore, necessarily two-dimensional and whilst it can illustrate the view of an object from a single viewpoint (like a photograph), it does not 'know' about the object as a three-dimensional entity. A scene or object can be saved from a modeller as a picture file, for instance, and can be subsequently loaded into a paint program where the bit mapped image (as it is called) can be manipulated. It can probably also be reloaded into a modelling program but only as a flat image to be used as a background or to wrap around an object for surface detail. It cannot be reloaded into a modelling program and manipulated again as an object since it is only a picture of the object. This rapidly becomes obvious through use but it is not uncommon to find a student new to modelling having abandoned a model in the belief that any file saved will contain all the model's information.

4.4 VECTOR

If the information about an object (or a shape) is stored in terms of the spatial relationships between its vertices then it is in vector form. A file with this information is needed for an object to be recreated in its own three-dimensional space. If a two-dimensional shape, such as a type font, exists in vector form, then the lines describing its outline are objects as opposed to being merely marks in a raster image. The precision with which a vector model can be displayed is dependent on the resolution of the current display device (e.g. a monitor screen or printer) regardless of the precision at which it is stored, which could be absolute. However, the accuracy with which it is stored determines the accuracy with which it can be mathematically manipulated by the program.

4.5 SPLINES

It is not easy to draw by hand a smooth line passing through set points but aids have evolved in other disciplines which can help. The 'French curve' is a template of curve profiles which draughtsmen utilise, and from shipbuilding comes a more complete solution. In order to draw the smooth curves of sections through ships' hulls, thin, flexible strips of wood or metal (called 'splines') were held down at key points by weights (called 'ducks'), and their natural, internal tension led them to take up a smooth curve through the weighted points. Mathematical equivalents of the shipbuilders' spline have been developed to provide us with a ready method of establishing a smooth path defined by a few controlling points. This is in contrast to a circle generated by a mathematical equation, for instance, whose smoothness depends on the number of points which have been used in its generation and display.

A curve can, therefore, be approximated by a raster display or by a continuous sequence of vector lines but can be only be accurately described mathematically. Bézier, working for Renault, developed one of the most commonly-known formulations in order to be able to describe the curved panels of car bodies. The Bezier curve is one of a number of different splines which are defined by an equation using

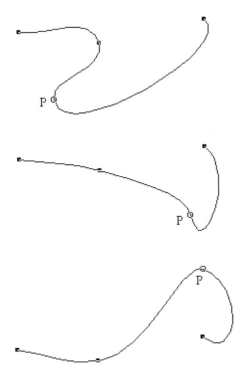

A spline curve modified by moving control point 'P'

control points to establish varying degrees of curvature along a line. Moving a control point changes the curvature, but the nature of the change varies according to the formulation used. Sometimes the change is local to the point moved, yet in other cases the whole of the curve is affected. It should be noted in passing, that most types of spline curve do not normally pass through their control points. Current modelling packages are tending to use NURBS (non-uniform rational b-splines) which are the most flexible for many purposes. A surface can be created from a net of splines, the Bezier patch being an example in which adjacent patches with the correct continuity combine to create a complete surface.

Splines are extremely valuable now that modelling has developed beyond simple geometric shapes and tries to deal with contoured surfaces. They enable the creation of three-dimensional objects and surfaces whose contours can be smoothly and easily changed by the movement of control points. Most modelling packages now support splines, though with different degrees of sophistication. The use of splines to define fonts allows many packages to create or import two-dimensional type for subsequent conversion to three-dimensional letter forms.

4.6 MODELLING

There are a number of ways of creating and describing 3-D objects such that their data can be stored and manipulated. Some methods log the history of the object's creation, some store a mathematical description and some list the position and connection of the object's vertices. Some deal with solid objects and some deal only with the surface facets of the object. Some descriptions accurately record the object but 'freeze' it so that the possibility of future manipulations becomes limited. All these methods have their own rationale and relevance, and most appear in current modelling packages, but their structure is often hidden from the user.

4.6.1 METHODS

Any object, even a simple cube, can be created within modelling packages in several ways. The significance of the method used might affect how that object can be treated or how it can be used in the process of creating further objects. For example, the data created in making a cylinder with the B-rep 'spin' method will be for a faceted object, whilst a cylinder in a CSG system will be understood as infinitely smooth. Since data types can be converted internally by the program, it is not necessarily obvious which method is being used. Fundamentally different principles for making objects are not, however, normally available within individual packages so it can sometimes be important to select the right modelling type for your particular requirement. This is not too difficult as products are often advertised as being for different markets, such as architecture or design visualisation, and for many unspecialised tasks any method can be adapted to serve.

4.6.1.1 B-REP

The most common method is to represent the boundary of the object, known informally as the 'B-rep' method. With B-rep the surface of an object is polygonised and the description stored as a list of vertices (the corners of the surface polygons), a list of lines joining the vertices (i.e. the edges of the polygons) and a list of faces identifying the individual polygons. For the purpose of rendering the object, these polygons are usually triangulated (divided into triangles) since triangles are necessarily planar and so unambiguous surfaces. Triangulation is not necessary to the description of the object. However an object is produced, it can be given a B-rep description although this might prove to sacrifice accuracy.

Eight typical primitives

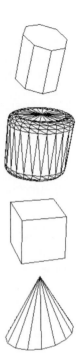

4.6.1.2 PRIMITIVES

Modelling packages usually have available a small library of simple, generic three-dimensional models called primitives. These can be represented internally as individual entities (rather than as polygonal descriptions) which use little memory. Typically these will be a cube, sphere, cylinder, cone, torus, wedge, plane and perhaps others, the ability for the user to add more sometimes being available. These primitives can be scaled and modified within the application, often both interactively (such as with a mouse) and by numerical input, so that a sphere can become an egg and a cube can become a customised parcel. These primitives provide building blocks for more complex objects (e.g. a simple camera being made from a cylinder abutting a cube).

4.6.1.3 SWEPT FORMS

There is some inconsistency about the use of the term 'swept form', or more correctly 'swept surface model'. It has started to be used to describe a freeform subset of the general class of swept surfaces, but will here be used to refer to the whole class. The class can be summarised as being created as the result of a two-dimensional (XY) section being 'swept' through the third (Z) dimension.

4.6.1.3.1 SPUN

A two-dimensional template, either a closed or open shape, can be rotated about an axis (conventionally around the Y axis) to create a B-rep form. The traditional example is the creation of a bottle and wineglass, or, if a circular template is located outside the centre of

38

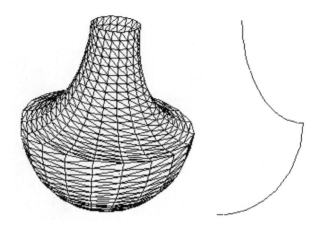

rotation a form like a doughnut is made. In these examples the arc through which the template must be rotated to make a complete form is 360°, but most modellers allow for partial rotation in order to create a form such as a melon slice. In the event of less than full rotation, it is necessary to decide whether the sections at either end of the sweep are to be solid surfaces ('capped') or whether the form is to be open ended ('uncapped'). This modelling method is called spinning or lathing.

It is often possible to rotate the template about a changing point. If the point of rotation moved in a straight line along the Y axis, whilst rotation was about the same axis, then a corkscrew form would result.

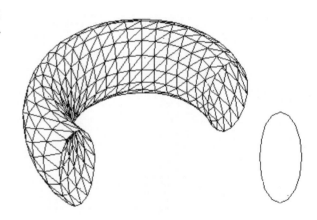

4.6.1.3.2 EXTRUDED

If a template is swept in a direction orthogonal to the plane in which it lies, the resultant form is described as extruded. As an example, a square section could be extruded to make a cube, or a type font could be extruded to make a three dimensional letter form. As with spinning, the ends can be capped or open, so an uncapped cube could be seen as a length of square section pipe. If the line of extrusion is not straight (typically a spline curve option is an

Two spun forms with their templates
Top: With axis of rotation in contact with the template
Bottom: With template distanced from axis of rotation

39

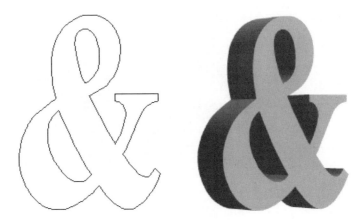

Template and form extruded from it

alternative) and / or if the template is also revolved, then complex twisting forms can result. The template may be held orthogonal to the path (like a pipe going around a bend) or else translated along the path (maintaining the same orientation).

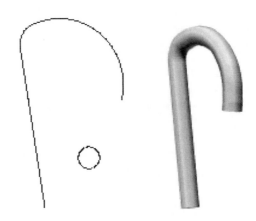

Circular template and path used to extrude form

4.6.1.3.3 FREEFORM

Freeform is another word that finds itself variously applied to
different aspects of modelling processes. Sometimes it is used to
refer to the quality of a surface or form which is asymmetrical about
all axes and sometimes to describe a surface or form which has a
loose, sinuous quality as if drawn freehand (as opposed to being
constructed with an obvious geometry). In some packages a specific
freeform tool is available, and this normally enables the construction
of an object by defining separate templates orthogonal to each of the
three axes.

4.6.1.4 LOFTED

In the lofting process, consecutive cross sections through an object
are joined by triangulation, a standard technique for creating an
optimal surface of triangular patches between the edges of the
sections. The cross-sections could be thought of as being like the
geographical contour lines that define a hill, and the triangular
patches as being the sides of the hill itself, the smoothness of the
surface depending on the closeness and detail of the sections. The
alignment of matching points on each section determines the
model's coherence and usually results from the order in which the
points are created, though in sections having different numbers of
points the modeller might not connect them in the way intended by
the operator. For this reason it is sometimes possible to specify
which points are to be aligned. Lofting would provide a good way
of constructing a ship's hull from its sectional members.

4.6.1.5 SKIN

The ability to construct the 'skeleton' of a form and then wrap a
surface skin around it to create an object is very useful. The skin
function available in some modellers appears to be merely lofting,
but in more sophisticated form a skin uses a scaffolding of defined
points at the intended surface of the object to create a splined
surface.

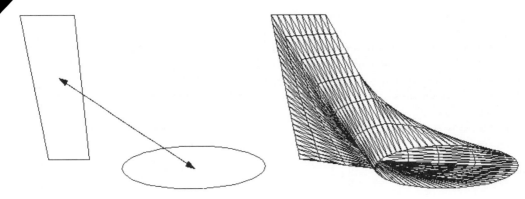

*Two end templates
used to create skinned
form*

4.6.1.6 CSG

A modelling method popular in CAD systems is that of constructive solid geometry (CSG). In this approach an object is represented as a combination of simple primitives such as cube, sphere and cylinder. These basic solids are used as building blocks for more complex objects by means of a system which uses mathematical descriptions of the spatial relationships the primitives have to one another. Boolean set operations of union (+), intersection (-) and difference (&) to describe the logical operations of adding two objects, subtracting one from another, or defining the overlap between two objects. The primitives can be scaled and it is also possible to define primitives by the use of 'half-spaces', which are infinite surfaces dividing three- dimensional space into solid or void to define objects (any point exists in either the solid, the void or on the division, and several half spaces can combine to define the space enclosing an object).

CSG is very economical in the information it stores, as the primitives are stored as mathematical forms rather than by listing all the vertices, etc. For example, it might require the storage of 45,000 coordinates alone to describe a moderately smooth B-rep sphere (in addition to all the necessary edge and face information) whilst a perfect sphere can be defined mathematically by four numbers – the three coordinates of the centre plus the radius. Complex objects are stored in a tree-like data structure that records the primitives used and the sequence of operations carried out on them.

Real world accuracy is one advantage of the system. This can be seen if a comparison is made between the CSG and B-rep methods for building a model having two rotating drums in contact with one another, such as might be required by an engineer. Because the CSG method uses mathematical descriptions of the primitives, the drums will have a constant diameter; however the B-rep method creates a polygonised approximation of a curve which means that the drums will always be faceted (however small the polygons used). This means that if the B-rep drums were brought together they would either have edges or facets in contact and would not be able to revolve freely. The comparison is somewhat confused by the fact that although a system is using CSG it will probably convert its models to B-rep in order to display them on screen for ease of rendering. (We will see later that it is possible to use raytracing to render directly from the mathematics that constitutes the CSG objects, but that this particular method is too time consuming to be used for any but the final display).

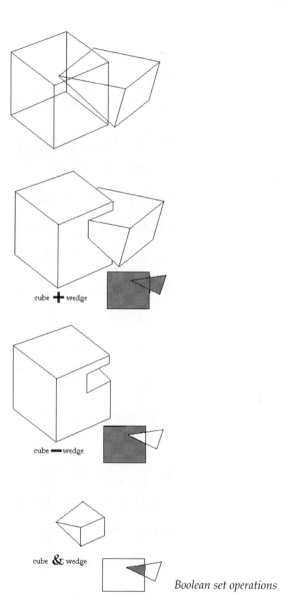

cube **+** wedge

cube **—** wedge

cube **&** wedge

Boolean set operations

4.6.1.7 VOXELS

A simple method which is of increasing interest, though currently

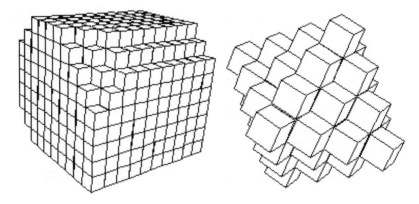

*Two forms made from
cubic voxels*

used only in specialised modellers (such as for medical use), is that of volume modelling, described here for the sake of completeness.

Correctly known as spatial occupancy enumeration, this method divides three-dimensional space into cubic units called voxels, of whatever size is suitable, and the object is described by recording the units it occupies. A voxel is sometimes described as a three-dimensional version of a pixel (although the former is really a unit of object space and the latter a unit of screen space). Because, although blindingly simple, the method currently requires extensive data storage in order to describe an object at a useful resolution, various storage and search techniques have to be used with it. Volume models are very easily rendered, and the ability to 'peel' the models back in layers can be most useful.

4.6.1.8 FRACTALS

Fractals are becoming increasingly used in modellers to simulate
structures like landscape where credible detail can be achieved by
fragmenting a surface or object in a pseudo-random manner without
recourse to a large data bank. The subject is too large to discuss here,

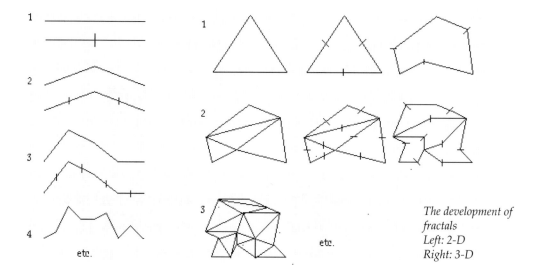

*The development of
fractals
Left: 2-D
Right: 3-D*

but its use to generate clouds, mountains, rust and such like is
finding increasing use in recent software.

4.6.1.9 PARTICLES

Particle systems are starting to become included in a number of
modellers but with their intended use limited to a few special types
of model. They are a particular favourite of mine and consist of a
large number of particles, typically between 10,000 and 1 Million.
Each particle represents a single point in three-dimensional space,
and as a group they can simulate 'fuzzy' phenomena such as clouds,
flames, grass or a waterfall. The particles are very easy to define,
create, colour and move and they have proved very useful in
scientific simulations for showing gas and water flow.

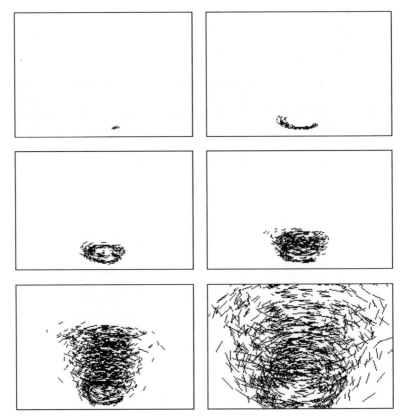

A 'typhoon' particle system where each particle is represented by a line showing its recent path

4.6.1.10 GRAMMAR-BASED

Once more outside the scope of most existing software, grammar-based systems use inbuilt rules to generate objects and have notably been used to simulate plant growth. The method is listed here as it presents a contrasting paradigm to the more conventional modellers, but though the structure of an object evolves differently using this method, it must still be made manifest through use of a previously-described method.

4.6.2 JOINTS

Not all the models we might want to build are made exclusively of separate, unrelated objects. Often objects have several connected parts, either physically jointed (like a door in a door frame) or in a fixed but distanced relationship (like a planet and its moons). Modellers normally allow these relationships to be established, firstly by creating links between objects (using a parent/child metaphor) and then by defining limitations to the freedom of movement of the links. It is thus possible to define a forearm as being the 'child' of an upper arm and the upper arm as the 'child' of the torso (which is its 'parent') in order to create an arm which has joints at elbow and shoulder. The two joints can have their movement limited around the X, Y and Z axes by set amounts to create a hinged elbow and a ball-jointed shoulder. When the parent object is moved its 'children' stay connected, but in any positional relationship permitted by the joint constraints.

4.6.3 FLEXIBILITY

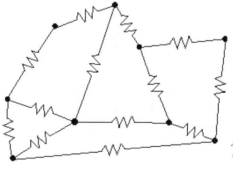

But what if the object we wish to create is a jelly (jello). We can build a model of the jelly in any position it might take but the internal flexibility of the form is only evident over time and so this is more of a problem to an

Springs and dampers forming the skeleton of a flexible object

animator. There are ways of defining variable distances between vertices (which can be connected by mathematical 'springs') but these are not available in most modellers. It is, however, possible to duplicate some symptoms of flexibility through the judicious use of scaling and editing functions in successive views of an object.

4.6.4 DEFORMATION

In high-end modellers there is sometimes a facility for transforming the shape of objects called freeform deformation (FFD), not yet universally available but likely to become so in the future. Instead of attempting to transform directly what might be a very complex object, the transformation is applied to a cage of control points which imprison the object, thus enabling correct relationships within the object to be maintained. It is as if the object were set in a block of flexible resin which could then be bent and modified, the captive object accepting the same transformations as the block.

4.6.5 EDITING FORMS

Basic modelling methods typically create even or symmetrical forms, which can be combined to create more complex models. This is often an inadequate or inconvenient means of creating the desired contours, and the ability to manipulate the object or its surface more intimately is highly desirable. Low level editing at the level of

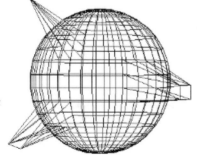

A spherical object edited by dragging a vertex (top), a facet (centre) and a line (bottom)

individual facets, lines and vertices is desirable and starting to become available. It enables the operator to create an approximation of the required object using traditional methods and then to reposition, rescale, remove or supplement existing details on the model. This manipulation can entail dragging a single vertex to a new location, moving a group of vertices, adding vertices (and correctly connecting them to the rest of the object) or carrying out similar operations on lines, facets or defined sections. Editing should also permit cutting planes to slice objects into pieces, holes to be bored through objects (though this might be carried out in the model construction phase) and vertices or defined parts to be removed. Adding and subtracting features, as opposed to just reorganising them, carries with it the more difficult problem of retaining the desired coherence of the object, since if a vertex is removed the existence of lines and facets that rely on it must be dealt with by either the operator or the system.

4.7 VIEWING MODELS

Having created a numerical description of the desired model it is necessary to display it on the monitor screen. Since we probably have a mental picture of the object, it might not be immediately obvious that we need to decide on a form in which to display it. In fact the process of creating the object is normally an interactive one in which a screen image is used and so the decision will probably be made before creation starts, but we need to be aware that there are many forms the image can take and that choice usually involves a compromise between time and legibility.

4.7.1 LEGIBILITY

The range of ways in which an object can be presented is quite consistent between packages. The criteria for selecting a viewing method vary at different stages of the modelling process, but the relatively long time taken by some methods is often a factor. In the building stage it is often necessary to have effectively instantaneous feedback on the screen for every change of position or scale of an object made. If, as will often happen, a mouse is being used to drag one object towards another, the operator relies on the screen updating in real time in order that the objects' relative positions can be continually assessed. During this operation a 'bounding box' is often automatically created. This is a heavily abbreviated representation of the model by a box corresponding to its maximum dimensions. Owing to its simplicity, its movement on screen can be in real time.

Later in the process the position of objects may be finalised but the effect of lighting on them may need to be seen and this requires a different process, the longer time taken being acceptable.

4.7.2 WIREFRAME

The simplest representation, and quickest to display, is a wireframe model in which all of the edges are shown as lines. This can be confusing to view, however, as we are able to see the back as well as the front of the object and this lends itself to the manifestation of optical illusions. Because it was the earliest form of representation of an object on a computer screen it is still sometimes called for when a scene is required to have a 'computer generated' feel to it.

Some improvement can be achieved by using intensity modulation to strengthen close lines and make distant lines fainter, but unless it is transparent, the front surfaces of a real object obscure the back surfaces, an important factor in our visual understanding of the object. (In fact, even a transparent object usually has the back surfaces modified in some way by being viewed through the front ones, either changed by a shift in colour or tone or by refraction). A wireframe view can only be constructed, of course, from a model which has been built from vertices, such as in the B-rep system, or else can be converted to such a system at this stage.

4.7.3 HIDDEN LINE

An improvement on a wireframe model is a hidden line version in which the front surfaces obscure those behind, resulting in a marked decline in ambiguity. This ability to create a more realistic view is so important that many algorithms have been created to do the job. The manner in which they work depends on the way in

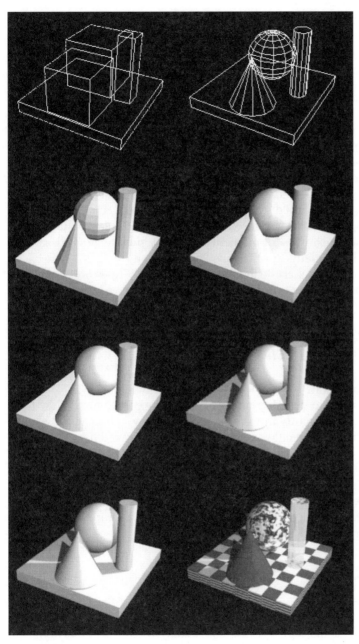

A model displayed using (from top to bottom, left to right): bounding boxes, wireframe, Lambert shading, Gourard shading, Phong shading, raytracing, raytracing with optimum smoothness, with added surface textures

which the model data is held and on the level of accuracy required. There is usually a trade-off between sophistication and speed, as a general purpose algorithm (which might not be able to deal with special cases) is likely to be much faster than one which is built to test for, and resolve, all conflicts it might meet. Applications sometimes provide a quick method for draft work, at which stage errors might be more acceptable than time delays, and a more efficient method for final work when the extra time taken is an acceptable overhead.

The simplest way of using hidden line removal to improve on a wireframe model is by back-face culling, in which surfaces pointing away from the viewer are removed. The direction in which a face is pointing is established by checking the angle between the viewer's line of sight and the 'surface normal', a perpendicular to the surface in question, using vector mathematics (both surface normals and vectors have many applications in computer graphics). The back-face cull is a rather crude method, however, as it does not deal with objects overlapping one another and usually proves inadequate for objects of any complexity.

4.7.4 DEPTH CUEING

As was mentioned when describing wireframe models, an improvement in legibility can be achieved by coding areas close to the viewer differently from those furthest away. This can be accomplished with either colour, perhaps having red near parts graduating away to blue distant parts, or with intensity, where near parts are displayed more brightly than those at the back. The calculations to achieve these ends slow things down and are not, therefore, normally used during model building, but can be useful at a final stage, particularly if a wireframe view is required. It is not, however, a feature often found in modelling packages.

4.7.5 SOLID

A more realistic impression of an object, and one which is far less likely to be ambiguous, is a representation in which the facets are shown as solid. This matches more closely our understanding of objects in the real world. It is in the nature of a solid model that hidden surfaces are obscured and a number of methods are used to achieve this effect.

Commonly used is the 'z-buffer' method in which the spatial depth of each surface is checked at each pixel location, and the closest surface (i.e. the one with the smallest Z value) is displayed. Another is the 'painter's algorithm' which displays the furthest surface and then works forward through space 'overpainting' with closer surfaces (although the distance of a surface can be ambiguous if it is not parallel to the viewing plane). It is inherent to some rendering methods, such as 'ray tracing', to solve the hidden surface problem.

4.7.5.1 CODED AND LIGHTED

Although the obvious way to show an object as solid might be to treat it as if its surfaces are illuminated by a light source within its own world, it is equally possible to shade facets systematically. If all facets facing in a given direction are coded in the same way, either with the same tone, pattern or colour, then a logical model is produced which has much of the visibility of a lit model. For example such a system might be employed to single out all surfaces of an architectural model facing North.

4.7.6 PERSPECTIVE

We see the real world in perspective and the ability to give our model a perspective projection strongly enhances its ability to be understood clearly. Most modellers have the ability to change the degree of perspective by altering the angle of view and distance from the model, usually using a camera metaphor. (It is worth noting the error of the commonly-held belief that changing the focal length of a camera lens changes the perspective of a view. It is the change of distance from the subject which the change of lens permits, that effects the perspective.) The perspective observed through a

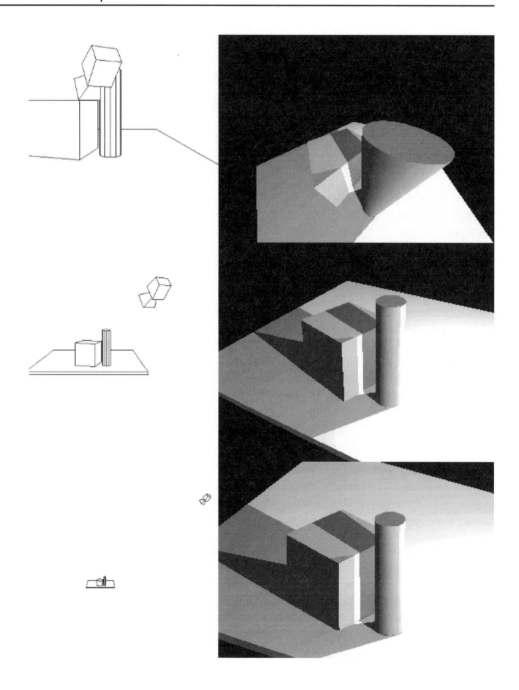

wide-angle lens when close to an object is dramatic, and that seen with a telephoto lens at a greater distance is much less pronounced.

However, there are occasions when traditional (single viewpoint) perspective can be a problem to us. When trying to set one object so that it is just resting on another, perspective can confuse the issue and modellers should allow you to select an axonometric projection in which perspective is eliminated and the task is made simple.

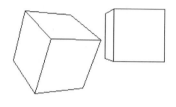

Previous page: Changing focal lengths in camera view to simulate wide-angle (top), normal (centre) and telephoto (bottom)

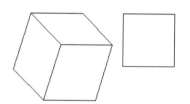

This page: Perspective projection (top) compared with axonometric projection (bottom)

CHAPTER 5

FUNDAMENTALS: RENDERING

Chapter 2 looked at the nature of rendering and the way in which it could simulate, with ever increasing realism, the conditions of the real world. It should be noted that packages often devise their own terminology to describe lighting and rendering methods but they can usually be recognised as similar to those types and methods listed here. Whilst modellers such as 3D Studio now have comprehensive rendering features built in to them, extra facilities may be found in a separate renderer to which models can be exported.

5.1 LIGHT SOURCES

There are several lighting types, available in even the simplest applications. A spotlight is a light source in which the beam spreads out from a specific point like a torch. The beam may be restricted in its arc (as if by a shade of some sort), the angle of the light beam being alterable in a special panel. It can also be subject to the 'inverse square law' in which the light intensity decreases in proportion to its distance from the source, and this fading can usually also be set in the lighting panel.

A point light simulates a source such as a light bulb which radiates light in all directions, and a variation which simulates the sun is sometimes available (i.e. the rays are given direction but are treated as parallel and of consistent strength).

A third light type is 'ambient', which is calculated to illuminate all surfaces with consistent strength and without direction and is often found useful in relieving the totally shadowed areas created by directional light sources in simpler lighting models.

5.2 LIGHTING MODELS

As you might expect, or else will soon find from painful experience, the lighting models available range from fast, crude renderings that are almost immediate, to slower, more subtle methods that can take hours, or even days, to complete. There is a constant trade off between subtlety, smoothness, resolution and speed. For this reason modellers usually let you work rapidly with a basic render but give you the ability to apply the top quality render to a selected part of the scene as a check before committing to a time-consuming full scene render. It is, therefore, possible to confirm details such as the extent of shadows or the degree of reflection, and to make necessary adjustments, at an early stage. The speed at which renderers work is tied to the hardware configuration in use but updates to modelling software often improve on the efficiency of algorithms used. This is particularly important as we soon take for granted the speed available to us and stretch our renderers to their limits.

5.2.1 LAMBERT

The simplest shading model is Lambert shading (sometimes labelled 'fast shading') which uses the cosine of the angle between the ray of light hitting the surface and the surface normal, to establish what the intensity of the surface should be (hence its alternative name of 'cosine shading'). As the light source comes to be closer to a perpendicular from the surface, so the angle decreases and the surface becomes lighter. When the light is at right angles to the surface the angle is zero and the light intensity is at maximum. Despite its simplicity, it adds enormously to our perceptual understanding of the object and, since it can be applied extremely quickly, is usually included as the basic shading method in applications. It does, however, produce flat shaded polygons which emphasise the artificiality of the model, and lacks any gradation across planes.

5.2.2 *GOURAUD*

Henri Gouraud gave his name to an improved shading model published in 1971 (sometimes labelled 'smooth shading'). It averages the light intensities at the edge of each polygon and then interpolates along each scan line across the plane lying between these averages to give a smooth, eggshell-like gradation. An extension of his model also allows the individual facets to be hidden by interpolation across facet edges.

5.2.3 *PHONG*

Two years later in 1973, a paper by Bui Tuong Phong introduced a method which added specular highlights to smooth shading. Phong shading, or 'highlighted shading', calculates the intensity at each point along a scan line from its approximated normal. (The approximation is arrived at by interpolating from the normals at the edges on that scan line, which are in turn interpolated from the normals at the points bounding that edge.) Surfaces with different levels of shininess can be simulated by employing a gloss parameter to determine the size of the highlight area. This 'specular' highlight depends on viewpoint, unlike a diffused surface which is independent of the eye position.

An odd effect of both the Gouraud and Phong methods is that an object, such as a polygonised sphere, can be beautifully smooth across its surface, but will still have a horizon made up of the straight polygon edges. This requires additional treatment which is unlikely to be available in a basic modeller, although the smoothness can be improved by increasing the number of facets in the model. It is also necessary to make sure that objects which are meant to look polygonal do not unintentionally have their edges smoothed over.

5.2.4 BLINN

Some modellers incorporate a shading model named after computer graphics guru James Blinn, which improves on the Phong model by approximating specific materials by the addition of some extra calculations (derived from real materials) and applications usually have a library of surface types, such as silver and brass, available. The characteristics of the library surfaces can normally be altered, with different degrees of subtlety, and new surfaces created.

5.2.5 RAY TRACING

A more recent method called ray tracing employs a different technique to great, and popular, effect. The principle is simple: a ray is traced back from the viewing position through each pixel to the first surface it meets. The ray is then reflected from this surface into the scene, reflecting off subsequent surfaces until it reaches a light source or leaves the scene. The pixel is then set according to the intensity and colour of the light remaining after the contribution of intervening surfaces. A lot of computation is required, increasing with resolution, and a limit must be set to the number of times a ray can be allowed to reflect before a final result is accepted. The more reflections each ray is allowed, the more accurate the result. The method produces arresting images and automatically deals with shadows with refraction and with transparency but is currently too slow to be practical in many situations unless fast hardware is available. The method is good at dealing with specular light but poor with diffuse light.

5.2.6 RADIOSITY

Even slower, though good at dealing with diffuse light, is the radiosity interchange method which developed with the field of architectural design in mind. It works from the assumption that the light energy striking a surface must equal the energy reflected, transmitted and absorbed. The first requirement is that the whole scene be divided up into surface patches (which may prove to be the way it has already been modelled) and each patch effectively treated as a secondary light source. Extensive calculations consider the effects of every one of those patches on every other and would be almost impossible to compute if it were not the case that most of the patch pairs will prove to have a nil relationship.

This method produces very credible subtlety within shadows and penumbra (soft-edged shadows), and has the advantage that the computations are independent of viewer position. This is very convenient for animation through the scene since the calculations need only be done once (provided that nothing within the scene changes during the sequence). Whilst convenient for animating movement through a fixed scene, it would be impossibly painstaking for moving anything within the scene, since all the calculations would have to be repeated for each frame.

5.2.7 PHOTO REALISTIC

Sometimes a renderer will have an 'ultimate' quality setting which gives the highest levels of refraction, smoothness, etc. intended for images destined for publication. The same qualities might be attainable by fine tuning other rendering models, but the default for this setting produces what the manufacturers consider the best image that their renderer can provide. The results can be spectacularly good but very time-consuming to produce.

5.2.8 SPECIALISED

New lighting models are being developed all the time to deal with some of the subtle and complex lighting situations that can arise in real life. By way of example, we can look briefly at two that have been presented recently. Mark Watt for example has used a variation of backward ray tracing (where the ray starts at the light source rather than at the eye) to render specular to diffuse phenomena such as the interaction of light with water. His method incorporates information about caustics, which deals with reflection and refraction by curved surfaces. With this technique he has produced some elegant and convincing animations of the delicate patterns that dance around on underwater surfaces. It is fascinating to see how evocative they are of other qualities we are all familiar with in swimming pools, recalling the memory of actually being in the water.

Nakamae looks at rendering road surfaces under various weather conditions, which has particular relevance to the development of driving simulators. His team at Hiroshima University has presented animations of road surfaces drying out, in which muddy puddles evaporate (requiring analysis of the minute undulations of the asphalt), but more exciting are those of cars driving at night. In order to simulate the effects of on-coming headlights he had to allow for diffraction due to the pupil of the human eye and even eyelashes. The results are uncannily effective and the most significant clue to the computer origination of the sequence lies not in the rendering but in the eerie smoothness of the car's motion.

Specialised lighting models such as these are not normally available in existing software but research such as this will find its way through to packages of the future.

5.3 SHADOWS

It is useful to have shadows dealt with automatically as happens with ray tracing. It is equally possible, however, to calculate shadows separately and a number of algorithms have been developed to do this. An architect might want to know how shadows are going to be cast by the sun without wanting all the other visual information that ray tracing would give him. It might also be useful to calculate the position of all shadows in a model (rather than those seen from one viewpoint) so that the view could be moved around a model without the shadows having to be recalculated for each position. Such shadowing is not likely to be found in general purpose modellers, but features such as the ability to create penumbra are sometimes included.

5.4 SURFACE CHARACTERISTICS

Since most users will not want all models to be rendered in a smooth, featureless monochrome, there are a number of methods available in most modellers to apply surface treatments to objects. These can usually be selected from a library included in the package which can be added to or edited at will. Colour, pattern, texture and images can be laid onto the surfaces of objects which can also be tuned for reflectivity, refraction, transparency and other qualities.

5.4.1 PROCEDURAL AND NON-PROCEDURAL

One useful distinction that can be made between different types of surfaces is whether the texture is defined by a mathematical function, which contains simple 'rules' by which any amount of the texture can be generated as needed, or whether the area of texture is stored complete. In the first case (procedural) there is very little to be stored by the application and the texture can be created indefinitely, optional random elements giving variety if required. In the second case (non-procedural) the whole texture map has to be stored in every detail. A chequerboard pattern can be generated procedurally, marbling can also be generated procedurally with a random element introduced, but a picture of a bunch of flowers has to be created elsewhere and stored.

5.4.2 TEXTURE MAPPING

Texture mapping is the process of transferring textural information from two-dimensional 'texture space' (where it is stored or generated) to three-dimensional 'object space' where it is applied to the required surface. The mathematics to do this vary in complexity depending on the recipient surface. Transferring a flat picture to the flat surface is relatively simple, wrapping it around a primitive with a simple geometry is somewhat more complicated, and applying it to a complex 'hand-made' object is potentially very difficult and time-consuming. Packages vary in their ability to deal with the more complex situations. Maps can be applied to surfaces in several ways, the method used being determined by the object's geometry and the effect required. For example, the map can be applied orthogonally (like wallpaper) or as if projected from a slide projector (when it is directional). It can also be 'shrink wrapped' around an object or wrapped as if with a rubber sheet.

5.4.2.1 DIRT

Most real world surfaces are not the unblemished paragons of perfection that the computer would like us to believe. In order to add another level of realism some renderers provide the ability to apply dirt to models. Rust can be generated with fractals, for example, and it is even possible to define the parts of the model most likely to be affected by rust and the degree to which each part will sustain the effect. It is only fair to say that this can produce very clean-looking dirt but its appearance is largely under the control of the operator.

5.4.3 SOLID TEXTURE

Instead of wrapping the surface of an object with texture, it is possible for the texture to run right through it. If an object is supposed to be made of wood, the solid texture method has the advantage that if the object is cut or opened up, coherent wood grain will be shown at the new surfaces. This method, which is associated in particular with CSG modelling, involves mapping from a three-dimensional texture space to the object space, which is a simple process. Such textures would normally be procedural.

5.4.4 BUMP MAPPING

Bump mapping (or perturbation mapping) creates three-dimensional texture by perturbing the surface normals using a mathematical function or a bump map. This means that the

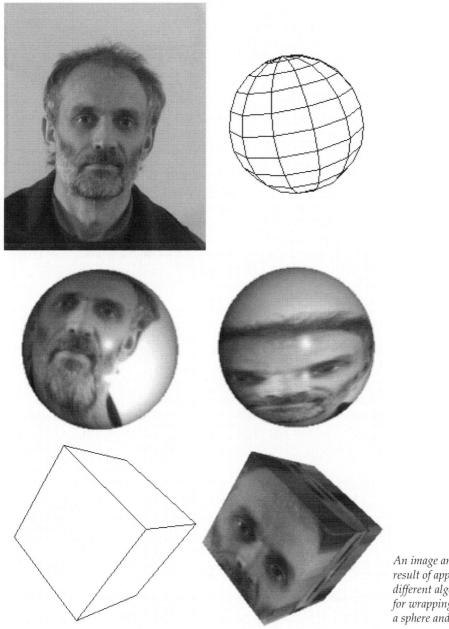

An image and the result of applying different algorithms for wrapping it around a sphere and a cube

information about the direction each facet on the model is facing (which is used by the renderer to set the lighting intensity on that facet) is disturbed, to fool the renderer into thinking that the facets point in a range of directions. This simulates a bumpy surface (orange peel is the standard example) and can be tuned to change the roughness, scale and uniformity of the texture.

5.4.5 IMAGE MAPPING

If you are a packaging designer, you will not be satisfied with creating a product with a uniform surface but will need to apply graphics to the box and a label to the product. This graphic image will be created outside of the modeller, in a paint program perhaps, imported and then applied by image mapping. Some control should be offered to enable the image to be scaled and aligned.

5.4.6 REFLECTION MAPPING

In order that a surface can be seen to be reflective, there must be something present to be reflected (for example, an object will not look convincingly metallic unless it reflects its surroundings). A surface can obviously reflect other surfaces in the scene but it is often useful to set up an environment map (an image which can be reflected in objects as if it were their world, typically clouds) or to wrap an image onto an object as if it were the reflection.

5.5 ARTEFACTS

A number of errors commonly arise in the production of computer-generated images, usually as a result of the fact that computers, and their display devices, work in discrete steps whilst the world we operate in is smoothly continuous. This results in our often having to match specific points in time and space to the nearest available points in computer time and space, with a small margin of error proving unavoidable. It might, for instance, be that the point in the model's world that we wish to display equates to an ideal screen location which is not exactly centred on a pixel. The best we can then do is to set the point at the nearest pixel, introducing a slight error. This type of error is described by a branch of mathematics known as 'sampling theory' and is particularly evident in CGI as spatial aliasing.

A straight line drawn on screen horizontally or vertically will appear perfectly straight since it will run along one row or column of pixels. A line drawn at 45 degrees will run diagonally through pixel locations lying in a straight line; but consider a line lying at an angle close to, but not at, those described. The requirement to match the desired line to the nearest pixel locations results in an uneven, stepped effect. As jagged edges ('the jaggies') can be quite destructive of the illusion we wish to create, much effort goes into removing them, or, more accurately, disguising them.

Anti-aliasing is a technique (which might be considered counter-intuitive) for disguising these effects by softening the edges of the line. Instead of representing the line with pixels entirely of the required intensity, the intensity of each pixel crossed by the line is set at a level between that of the line and that of the background, in proportion to the percentage of the pixel covered by the line. Where the line coincides with a pixel exactly it takes the line intensity; where the line crosses the boundary between two pixels they are both set at an intensity half-way between the line intensity and the background intensity. Anti-aliasing is difficult to achieve with a small palette but has become a standard option in most rendering software. Other methods of dealing with the problem are the 'dither matrix', which changes the intensity of a pixel on each scan sequence, and 'pixel phasing', where the screen location of individual pixels can be shifted fractionally by automatic adjustment of the electron beam; but these solutions are less likely to be encountered.

'Temporal aliasing' is a manifestation of the same problem in time rather than space and is relevant to animation. An example can be found in Westerns when the wheels of the stagecoach appear to be turning in the wrong direction.

'Mach banding' is a phenomenon, particularly associated with Lambert shading, in which a surface that should be smoothly shaded

Anti-aliasing applied to the lower half of a letterform

appears to have dark streaks on it. This anomaly is a product of our edge detection abilities and is most easily improved upon by decreasing the size of the polygons in use.

'Illegal colours' can be a problem at a production level, since it is possible to generate colours on screen which cannot be accurately recorded onto video tape or broadcast. A wave form device monitor, which is like an oscilloscope, can be used to spot the offending colours.

'Precision errors' are inherent to digital computers. The computer can only allocate a limited amount of memory to each number it uses, and if this space is insufficient to store the complete number it becomes truncated. The imprecisions created can accumulate to create noticeable errors, but is unlikely to present a problem for us in well-written software. It is interesting to notice, however, that in one application I have used, if 10 inches are converted to millimetres and then straight back to inches again, the result is not exactly 10 inches. (In another context, these rounding errors have been blamed, in the past, for false nuclear alerts!)

5.6 CURRENT RESEARCH

A lot of work is being done on new rendering algorithms, often in an attempt to match lighting situations found in particular environments in the real world. Although these algorithms may be designed for use within limited disciplines, and thus are not general enough to be of direct interest to us, the research finds itself taken up elsewhere and will lead to more sophisticated renderers, first on specialised high-end machines and later on PCs. The additional subtlety of new algorithms usually brings high computational overheads, which require a powerful machine to be able to handle in a reasonable time. Papers detailing the latest work are to be found in the proceedings of the annual SIGGRAPH conference.

CHAPTER 6

FUNDAMENTALS: ANIMATION

There is a large and growing number of disciplines in which animation is now a commonplace tool. As model builders we can use movement through and past our scene to add information which a static view will not offer. In real life it is often the case that an object is ambiguous until either it moves or you move relative to it, thus by animating either the viewing point or the object a clearer understanding of the model can be gained. In areas such as architecture, the ability to move around inside a computer model of a proposed building is increasingly sought and the feedback from such an exercise can lead to modifications of the original design.

A continuing problem with computer animation is that it looks like computer animation. Movements often have a queasy, mathematical smoothness which is unworldly. This is good for showing spacecraft, and perhaps acceptable for flying logos (since they are not of the real world), but is often poor for simulating real life objects.

It should be noted that changes in an animation need not only be spatial, i.e. moving an object or light to a new position, but can be enacted on any variable. The brightness or colour of a light can change over time, the colour or texture of an object can change, the position or focal length of a camera can change and the form of an object itself can change. It might prove very instructive to change only the position, colour and brightness of the light source in an architectural model to simulate natural lighting through the course of a day. 'Keyframe' and 'timeline' are the two main methods for controlling animation in current systems.

6.1 FRAME RATE

An animation is a sequence of separate images, called frames, which are presented in sufficiently quick succession to give the appearance of movement. Rather than set every individual frame (25 to 30 frames per second are used for a smooth animation) the operator will normally set important frames (known as 'keyframes') and leave it to the computer to calculate the frames in between these. It is still usually the case that the system will not be powerful enough to render the scene for each frame in 1/25th or 1/30th of a second and will therefore need to save it frame by frame before playing it back at full frame rate. The frames might be saved to video, in which case a single-framing VTR and an animation controller are required, or to the hard drive. To play back from the hard drive, particularly at full-screen, in full-colour, a fast-access AV drive is desirable. If the system is able to record and/or display the animation at the full frame rate the experience is described as 'real time'.

6.2 KEYFRAME

As mentioned above, a keyframe is a frame from a sequence at which point a significant event, such as a change of direction, takes place. In such a system everything in the scene is set in its correct position at each keyframe and the system generates the required number of frames inbetween the keyframes. The keyframes can be returned to and edited. In the simplest case of an object moving in a straight line between two points, the keyframes would be the start and finish frames. If the object was to move to this new position and then revolve 90° about its centre, the three keyframes would be the opening frame, the frame in which the traverse stopped and the end frame showing the object after it had finished its rotation. A high degree of subtlety can be used when defining the movement between keyframes, often using splines for smooth sweeping movement and 'cushioning' for measured acceleration and deceleration at the start and end of an object's travel.

6.3 TIMELINE

An alternative control method is to define the position of each object (which can include the camera or lights) at any given moment in the sequence. In this case each object in the scene is represented on the sequence control panel against a horizontal line divided into time increments. The position (or other attributes) of each object at any moment in the scene can be set on its timeline and is clearly seen relative to any other object. It is easy with this method to make adjustments to individual elements in a scene without affecting others and to make changes to the relative time taken by elements to enact changes. This system is sometimes extended to include timelines for a soundtrack.

6.4 SCRIPTED

The keyframe and timeline systems described allow the objects in a scene to be positioned on screen using the same controls that we are familiar with from modelling. An alternative is to write a script for the required movement, which is less intuitive but sometimes clearer, for example typing in a 45° rotation around the X-axis can be more accurate than making the rotation by hand. If the appearance of the scene is our overriding interest then we need to see it on the screen, but if the concept of the changes is more important, then a script might be simpler. I believe that the best animation systems provide control at both levels, but the facility is not always available within modellers. This is not unreasonable, however, as it is asking a lot of a package to have every possible feature of both modeller, renderer and animator.

6.5 KINEMATICS

Kinematics is the study of movement without regard to cause. At the time of writing most animation systems are kinematic, which means that it is up to the animator to deal with the degree and consequences of all movement within the scene. The acceleration of a falling ball, the amount it squashes when it hits the ground and the height to which it bounces must all be defined using the experience and intuition of the animator.

6.5.1 INVERSE KINEMATICS

It is often necessary for objects within an animation to arrive at a particular state in the final frame, or for an object to reach a particular interim goal (such as a figure sitting down on a chair). Rather than progress forward through the scene in the hope of arriving at the desired states it can be easier to work backwards from the end state and this can be achieved with inverse kinematics.

6.6 DYNAMIC

Dynamics is the branch of mechanics which deals with the way masses move under the influence of forces and torques. In a dynamic animation the system 'knows' the mass of the ball, the force with which it is started in motion, gravity and the ball's flexibility, so it is able to calculate the entire bounce sequence from the starting conditions that it is given. Such a system can be built to recognise collisions and to cause objects to make the correct response to them (which might be impossible in a complex kinematically-driven scene).

Dynamic systems are the object of much current research (and are fascinating to use) but the calculations involved can take a very long time and they are not appropriate in all situations. If the ball has to bounce across the screen in a commercial, arriving at a prescribed spot and taking exactly three seconds to do so, it can require a lot of fine tuning of the forces in the scene to achieve the desired end.

Systems which combine comprehensive kinematic and dynamic control are available but neither they, nor full dynamic systems, are as yet widely available.

6.6.1 INVERSE DYNAMICS

It is equally possible with a dynamic system to work backwards from a desired end state to determine the forces that need to be applied at the start of the animation. This is particularly useful in a dynamic system which needs to reach a predetermined end state in a set time, for instance a TV advertisement might require a snooker ball to bounce off several cushions and roll up snugly against a logo in exactly twenty seconds.

6.7 RULE BASED

The movement of elements within an animation can also be controlled by rules, although this usually requires some programming skill. Flocks of birds and shoals of fish have been simulated using simple sets of rules which condition the movement of the individuals, i.e. keeping close to neighbours, avoiding bumping into them, maintaining an appropriate velocity, etc. Such behavioural systems use artificial intelligence techniques and can be very effective in removing the need for the animator to hold responsibility for every detail of every movement at every moment in time. In the case of flocking, for example, the flock itself can be directed rather than individual birds within it.

Factors like gait cycles can also be defined in order to animate walking without having to set every step individually (a choice of gait patterns is often pre-defined in the system) and similar repetitive actions can be readily set to be handled by the system.

6.8 ASSISTANCE

Systems now commonly include ready-made movement types as an aid to animation. 'Explosion' might, for example, be available as a movement that can be applied to an object, causing its constituent parts to split apart and spin away from its centre. 'Banking' is another aid commonly provided which will tilt an object, or camera, as it travels around a curved path. A camera's view can also often be locked to an object so that it will follow the object wherever it travels and an object's orientation can be aligned to its direction of movement. Cushioning is usually available and unusual geometrical paths, such as a corkscrew, may be on offer. All these details make life easier for the animator though they can lead to choreographic overload.

6.9 HIERARCHIES

The creation of heirarchies of parts within an object has been explained in the modelling section, the standard example being foot connected to lower leg connected to upper leg, etc. Such a hierarchy will allow setting of the foot position to control correct positioning of the rest of the leg and body. This 'parent-child' relationship can also be exploited to control other types of movement through the creation of invisible parents. For example a child object could be set to rotate around its invisible parent and the parent then taken along a path to simplify a potentially complex movement pattern for the child.

6.10 METAMORPHOSIS

As well as being able to change the position, colour and surface qualities of an object over time, its shape can also be modified. The change from one form to another is a metamorphosis which can be easily achieved, particularly if the two forms of the object have the same number of vertices. If the initial object is established at some point in the sequence and then altered for an appearance later in the sequence, the system can often incrementally create changing objects at stages between the two extremes.

It is necessary in most systems for the object to retain vertex continuity during the metamorphosis since the system will be tracking the changes in position of the vertices. This rules out both editing that removes parts of the object and also metamorphosing between completely different objects. Some sophisticated systems can invent extra vertices in order to achieve a vertex match, though it is usually desireable to pair up vertices in the two objects by hand in order to control the quality of the change. An inventive operator will, however, usually find a way to achieve the required effect and producing animation is often about cheating.

It should be noted that the 'morphing' which is currently fashionable in TV advertisements is often achieved in two dimensions not three, by interpolating between still or moving images of the objects, not between the objects themselves. Software is now readily available to do just this and can sometimes provide a computationally cheaper way of creating a desired effect.

6.11 ORGANISATION

Unfortunately animation is often a gratuitous afterthough born of the ability of the system to do it. There needs to be a reason for things to move and planning for the way in which they move. The normal sequence of events in a project is likely to be construction, lighting and then choreography.

It is possible to improvise at the computer keyboard but little ambition is likely to arise this way. Animation is often best planned away from the computer using a storyboard – a sequence of pictures illustrating key moments in the script. Even if this is a simple, diagramatic cartoon strip it will often save much time and frustration at a later stage where conflicts between camera and object movement, for instance, can easily arise. Storyboards are also a standard means of showing a client what is proposed and can be preferable to animatics.

It is also wise to start off being sparing with movement as overuse can be visually disorientating. Analysing the many examples of good animation presented on television will quickly show that often only one thing is moving at a time and that movements are often most effective when subtle.

CHAPTER 7

PRACTICAL ISSUES FOR 3-D WORK

This chapter introduces the basic components of a computer system with particular thought for its use in the context of modelling and rendering. It does not refer specifically to any one PC configuration. Since the inexperienced user is catered for it is anticipated that some readers will choose to pass the chapter by.

7.1 THE OPERATING SYSTEM

As a user you are protected from the need to deal directly with the electronics in your computer by the operating system. The operating system can be accessed through a GUI (Graphical User Interface) such as 'Windows' that allows you to talk to the machine easily. This offers a 'desktop' on your screen when the machine is switched on, using windows, icons and menus that can be accessed by a Pointer, usually controlled by a mouse, and being known as a WIMP environment. Many details of the interface (such as desktop colour) and your interaction with it (such as mouse sensitivity) can be configured by the user.

The friendliness of GUIs has won many admirers and it is now the most common interface for personal computers, largely replacing the command-line interfaces which preceded it. There are situations where the old interface is more useful and sometimes more efficient (since the GUI can slow operation down) but the convenience and accessibility of WIMP environments has broken down much resistance to computing in general.

7.2 HARDWARE ELEMENTS

'Hardware' describes the physical components of a computer system; the machinery itself (as opposed to 'software' which instructs the hardware what to do). The whirring box on, or under, your desk, the monitor, the keyboard and the mouse are the basic hardware components. The vast number of PCs available offers a bewildering range of configurations but it is relatively easy to achieve a satisfactory compromise between cost and functionality. The legendary traps and pitfalls which used to cause problems for the unwary buyer in the PC market have been eased as compatability has improved.

7.2.1 CPU

The CPU is the heart of the computer, where the instructions specified by the program are carried out and where the operation of all other elements of the computing process is coordinated. It is built as a tiny integrated circuit (IC), often referred to as a chip, and has shrunk so much in the last 30 years that a computer as powerful as a standard PC would then have filled a room. There are several families of chips which are commonly used and each has particular features which cause manufacturers to build machines around it, VLSI technology (Very Large Scale Integration) now permitting the fabrication of millions of transistors on a single chip. A number of machines now use RISC technology (Reduced Instruction Set Computers). A RISC chip is optimised to run the bare minimum number of instructions very fast, concentrating on those used most often; infrequently-used instructions are constructed from combinations of the few present.

The speed at which operations are carried out by the CPU is determined by the 'clock rate', measured in megahertz (MHz), a megahertz representing 1 million cycles per second. Clock rates on basic PCs have risen in the last few years from around 25 MHz to 350MHz and are higher in each month's magazines. The faster the clock rate the faster information is processed, but since different machines do a different amount of work in one cycle it is not sufficient merely to compare clock rates. Bottlenecks elsewhere in the system can also serve to undermine the raw speed of the chip, a 'bus' (which transfers data between the CPU and the rest of the machine) being a typical example. The bus problem can be compared to a country lane which becomes inadequate when the villages it connects grow into cities.

Bench marks are used to compare the speed of different machines and to time each machine as it completes the same tasks, but whilst these tests are accurate they are not necessarily helpful. It might be that the tasks accomplished in the bench test are not relevant to the application in mind, that the machine is optimised to produce good bench test results, or even that the machine is too clever for the tests. The speed of operation is often described in 'mips' (millions of instructions per second) and 'mega-flops' (millions of floating point operations per second), with alternative prefixes 'giga-' (one billion) and 'tera-' (one trillion) increasingly being used to whet the appetite. For graphics applications it is often more useful to know how many polygon fills a machine can achieve in a second or how many vectors it can draw in a second, a calculation directly related to the end product and which takes into account any special graphics hardware aboard.

The basic chip is often supplemented by others which do specialised jobs and take part of the workload from the CPU. The most common is the maths co-processor, which relieves the CPU of much of the burden of mathematical calculation and is often now built into the CPU or else available as an optional extra. As modellers we might be particularly interested in graphics co-processors which take over graphics chores. These supplementary chips are optimised to undertake their limited functions more efficiently than any general purpose chip and allow the whole process to be markedly speeded up.

It is, unfortunately, necessary to acknowledge that modelling and rendering require the computer to do a lot of calculations and therefore are best performed on a powerful machine. High quality rendering, in particular, is too time-consuming to be considered in a production environment on the smaller machines, and indeed the software will often require that a co-processor is installed. The simpler modellers will run on any machine, however, and the area can be thoroughly explored without resorting to the most expensive hardware; you just need to accept that things proceed more slowly.

7.2.2 MEMORY

Computer memory is where data is stored, either permanently or while calculations on it are carried out. Its size is usually described in kilobytes and megabytes ('K' and 'Mb'), which stand respectively for thousands and millions of bytes (though strictly the value of a kilobyte is 210 or 1,024 bytes and a megabyte is 220 or 1,084,576 bytes). Sometimes memory capacity is described in words rather than bytes, a word being a unit of memory storage which can vary from machine to machine. A typical word on a personal computer is 8 bits.

Most memory is random access memory (RAM) in which memory locations can be written to and read from without having to work through a sequence of storage locations, and its contents are normally volatile (disappearing when the machine is switched off). Read only memory (ROM) is non-volatile, can only be read from (as a protection from being overwritten), and is therefore typically used to hold information such as the operating system.

Only a few years ago a typical home computer might have had 64K of RAM, but now 16Mb is a common minimum and 64Mb a good starting point for professional applications. The cost of producing RAM has fluctuated markedly and has had an obvious knock-on effect on the price of machines. At the time of writing RAM prices are low and it is realistic to expect a good memory size, but as this becomes the norm so new software will exploit the capacity and push demands ever upward. Top end machines, particularly in environments using graphics, might choose to start with 256MB installed and we will soon be measuring RAM in giga-bytes. Modelling and rendering programs can require at least 16Mb

of RAM as a working minimum and work far more comfortably with
four times that amount. RAM chips can, however, be easily
purchased and installed as your requirements grow.

Virtual memory can be created by designating secondary storage
to be treated as if it were RAM (and this is sometimes done
internally by an application for its own use) but it is slow and best
treated as an occasional stopgap only.

7.2.3 STORAGE

Since the information in the computer's memory will be lost when
you switch it off (or it crashes!), it is necessary to store it in a more
permanent form. You MUST, repeat MUST, acquire the habit of
saving current files at regular intervals lest an error (or a power
failure) wipes out hours, days or weeks of work. As well as saving
to your hard disc you must also back up your important files to a
secondary storage device in case you should have a problem with
your hard disc. You will accumulate much information (text files,
picture files, model files, etc.) that needs to be stored for future
reference, and the file you lose is always the vital one. The backup
storage device can be a removable medium or a 'mirrored' hard
drive, in which case all data is simultaneously stored on your main
and mirror drive.

Secondary storage is most commonly provided by disc drives
which hold information on magnetic 'floppy' discs loaded into the
drive, the standard currently being 1.44 megabyte of storage capacity
on a 3.5 inch disc. The term 'floppy', incidentally, describes the
nature of the magnetic disc itself which is now housed in a rigid
plastic case (unlike the previous generation of 5.25 inch discs which
were in flexible casing). Hard drives use permanently mounted
discs with very fast access times (now typically holding at least 250
Mb but cheaply available up to several gigabytes) and can be
mounted inside the main computer box or bought separately and
plugged in externally. All disc-based media allow random access,
which means that data can be accessed anywhere on the disc without
having to run through all the preceding material.

Removable storage media, in which a fat cassette slots into a drive mechanism, provide a convenient means of storing data typically in 44Mb to 750Mb units, and these are essential for passing files to computer bureaux. Those using the SyQuest mechanism are the most popular, but the Bernoulli version is often considered to be more reliable, though more expensive. Large removable drives of several gigabytes are now available and cheap removable storage devices are now often built into new computers as standard. Tape provides an alternative magnetic storage medium, but being a linear access method, is too slow for use in an application (imagine having to go through a dictionary from page one every time you want to find a word!) and is mainly used for relatively cheap back-up and archiving. DAT (digital tape) drives use this technology. Zip and Jaz drives have recently become popular largely due to their relative cheapness and easy portability.

Laser technology has dramatically increased memory storage capacity and is just starting to become commonly available for computers, having already established itself in CD players for HiFi systems. Standard rewritable magneto-optical (MO) drives hold from 128Mb on a removable 3.5 inch disc and larger optical drives hold 650Mb. 'Juke boxes' of discs provide even greater capacity. These discs provide more stable archive storage than magnetic material, are competitively priced and are becoming common removable media on the current generation of computers.

The magneto-optical 'floptical' is available to read both standard floppies and 21M discs but seems to have made little impact on the market. Data is stored magnetically but optical techniques are used to increase density and permanence. CD-ROM drives are now standard in most computers as a read-only medium for disseminating up to 600Mb of data and CD-ROM 'burners' now allow recording onto CD at an affordable price. The medium is particularly suited for applications using video/animation data which require a large amount of storage with fairly quick access times and is also a very stable archiving medium. It is now being used for holding applications accompanied by many examples and demonstrations and also for holding font libraries. CD-ROM drives are quoted as being 2x, 4x, 16x speed, etc. which refers to their increased speed (relative to the original drives) of access and transfer of data to the host machine, and software incorporating video often requires at least a 4x drive to avoid jerkiness.

Fast solid state secondary storage is also becoming used and its progressive availability is likely to be inversely proportional to the price of the material. Since the CPU can handle information faster than it can take it from secondary storage, slow access times can cause processing bottlenecks, which becomes a very real concern, for instance in animation.

The main factors governing the choice of secondary storage are: your typical file size and the number of files to be stored; the speed with which you need to be able to access the data; the security of the data; the cost of the medium and, in the case of removable media, the cost of the mechanism itself. Prices are currently dropping for all secondary storage devices, and in many cases, dropping fast. From personal experience I know that a 2Gb hard drive is now a tenth of the price that an 80Mb drive was just a few years ago. The cost of the devices themselves varies but those with the fastest access times and with extra features will normally cost more. If used for storage only this is less of an issue than archival stability. A rough comparison of the cost of the most common removable media (as at early 1998) shows that they have all fallen to between 0.15p and 0.35p per Mb.

7.2.4 I/O

As mere mortals we cannot commune directly with the computer and need means of getting information in and out. The PC is provided with a good range of ports (manifested as sockets) for the attachment of input and output devices. These devices are described in Chapter 8 and include hardware such as the keyboard for input and the printer for output. Different PCs offer different ranges of ports but many are common to all machines.

The SCSII port (pronounced 'scuzzy') connects external devices such as scanners and external drives and allows a number of devices to be 'daisy-chained' (meaning connected one to another in a chain). SCSII plugs/sockets are 25 or 50 pin, and the first and last devices in a SCSII chain need to be 'terminated', the first terminator often being built in and the last having a plug-in terminator. (I've never been aware of any problems with an unterminated SCSII chain but, apparently, I've just been lucky.)

The serial port connects to printers, modems and to a network.

The video port connects to a monitor.

The power ports accept power from the supply and provide power for devices such as the monitor.

The sound ports connect to microphone and speakers, etc.

The external disc drive port connects to an external disc drive.

7.2.5 NETWORKING

Networking is too specialised a subject to be dealt with in detail in this book but the basic principle is straightforward and very useful. PCs normally require an additional card to allow them to be networked but they can then be connected to one another allowing them to communicate with one another and with other devices on the network. This means that files can be transferred between machines, machines can share printers and other input and output devices, and common files can be held on a file server and accessed by all machines on the network.

Of particular interest to us as modellers, often wanting maximum processing power, is the ability of one machine to call on all the computing power on the network in the service of its own calculations. In an office with twenty networked PCs, this means that a single machine could be left running through the night with the power of all twenty machines at its disposal. Software is increasingly 'network aware', meaning that the software is designed to make use of this facility and can divide up the workload among available machines for optimum processing speed. Network rendering is becoming a common, and welcome, feature in modelling packages, often being made available through supplementary software.

Machines being used by the controlling computer can still be used for other jobs as their CPUs will only be engaged by the controller if they are idle. During word processing, for example, the CPU is heavily under-used and is therefore largely available for use by others on the network. Another side of network awareness is that software will now often check to see if illegal versions of itself are being run on the network and will refuse to function if that is the case.

7.3 EXPANSION

PCs have slots to take expansion cards for video and graphics applications, networking and communications, additional processing power, and other purposes. These cards can still be foot-long, printed circuit boards but are increasingly using the much smaller PCI standard. It is well worth trying to anticipate future expansion requirements when buying your machine and ensuring that you have a machine with enough slots to allow it to grow with you. If you want to squeeze maximum speed from your machine, and work with video and use a large colour monitor, you will need enough slots available for all the extra boards you are going to accumulate.

7.3.1 ACCELERATORS

I work on a machine that was state-of-the-art five years ago and used to delight with its speed of operation. Briefly. It is a well-known law of computing that the machine you need is the one that is just a little more powerful than the one you use, and this law is never clearer than when working with graphics, especially rendering. Since a high resolution, full colour, ray traced scene can still take hours to render, this impatience is, perhaps, understandable. However rapidly algorithms and hardware are improved, they are likely to lag behind demand, and only interactive, photo realistic, real-time image production is likely to satisfy.

In pursuit of these ends, a number of manufacturers make cards that speed up aspects of your machine's operation. At their most complex they can seem like computers in their own right (and can cost as much), whilst there are more modest improvements to be obtained at lesser prices. These cards are often sold on the basis that they will make your machine run faster than a model higher up the product range, and it is always worth checking that the cost of doing this is not greater than trading in and buying the machine whose speed it will emulate.

General purpose accelerators are available with various specifications, and in some cases can be added either singly or in groups. When grouped, dedicated software can then arrange for each card to be treated as an independent computer, allowing tasks to be distributed among all the cards.

More specialised cards aim to improve the machine's performance in one specific area, such as speeding up the operation of Adobe Photoshop (a widely-used photo design and production tool). Some modellers/renderers can be supplied with their own boards and multiple boards can sometimes be added for very fast processing. Increasingly, modellers and renderers are designed to make use of accelerator boards, if installed. The financial cost can easily be worthwhile in a production environment where time is money but is often outside the range of the casual user.

7.4 CONTROL HARDWARE

If the process you are working on involves machines other than the computer, it can be convenient, even essential, to have the computer control them. When producing images to be laid down on video tape, it is often necessary for the computer to spend minutes (or hours) generating each frame (video runs at 25 frames per second in the UK and at 30 frames per second in the USA). An animation controller, which can be a NuBus board like the DQ-ANIMAQ, will control a suitable, professional standard VTR (video tape recorder) so that it advances a single frame and receives each image as it is generated. It can also control a VTR which is inputting single frames to the computer, possibly in order to post-process or composite images. Animation controllers can also be external devices or software. Animated material is increasingly accessible in real time from fast AV speed hard drives and then needs only to be dumped to video on completion, which removes the need for separate device controllers.

7.5 FILES

When you save your work, entering a name in the file-save dialogue box, the data is stored in a particular structure or file format under that name. The structure used will be determined by the application, which may offer you a choice of formats. Sometimes these formats have been developed for the particular application alone and sometimes they are generic, in which case they will be encountered in many other applications. Generic formats obviously allow you to open files within different applications and thus take advantage of their different features, but specific formats are sometimes created to allow the storage of special information relating to just one application. For instance, a modelling file might include the state of the background, lighting and animation script that relates to its own application rather than just numerical details of the model which could be universally understood. For that reason, if a generic format is used for the sake of portability, it can be sensible also to store files in the native format so that all the information is retained for future use.

Some file formats originated with early applications in the field and have been adopted by later programmers; others provide updates to previous file formats. We might distinguish between the requirements of files containing words, model data and 2-D images, although all are stored physically in the same way. Many applications have their own proprietary file type, some having proved sufficiently useful to be supported by other applications. It is important, when considering new software, to ensure that it can import and export the file types that it will need in order to work with your other applications (and with outside applications if necessary). File conversion utilities can bridge most gaps and are an important part of a professional's armoury, particularly if they save time with batch-processing facilities.

7.6 COMPRESSION

It is possible to maximise the use of storage space by compressing information when it is stored and then decompressing it when it is retrieved. For graphics applications, one commonly used compaction technique is 'run-length' encoding. Rather than separately store the actual intensity of every pixel, run-length encoding stores the intensity of a pixel and the number of following pixels with that same intensity. Imagine a single pixel being ON at the centre of a 640 x 400 pixel display. Instead of individually recording the state of all 256,000 pixels, it would be sufficient to record that the first 127,999 pixels were OFF, the next one was ON and the remaining 128,000 were OFF. The efficiency of the technique is greatest in images with blocks of similarly set pixels and would become inefficient in the rare case that no pixel was the same intensity as its neighbour. It is, therefore, more efficient in dealing with computer-generated images (where the pixels in a solid block of colour are consistent) than in dealing with video images (where visual 'noise' limits the likelihood of adjacent pixels being the same). A number of other methods for compression are available and some applications can select from a library of different techniques after assessing which is most efficient for each given image. The encoding and decoding can sometimes be achieved in real time using either software or dedicated hardware methods.

Animation, and the increasing use of video, highlight the problem. If you consider the amount of storage required to store one second of high-resolution, 24-bit colour animation, it is clear that efficient compression algorithms are essential if we are to be able to develop the medium. Even at 640x480 pixels, one second of 24-bit, 30 fps video requires 30 MB of memory, and one minute needs 1.8 gigabytes. Assuming the storage is available, there remains the problem of getting that much data to the screen fast enough for real-time display. A common method of compacting animation is to store the first frame in its entirety and, thereafter, to store only the changes between subsequent frames. This is extremely efficient when few pixels have changed between frames, which may often be the case in computer-generated sequences. It is likely to be far less efficient in handling sequences from a video source, since random changes are likely to occur to pixels throughout a sequence, even in areas of apparently unchanging colour.

Compression algorithms are described as 'lossy' or 'lossless' according to whether they achieve their compression at the cost of sacrificing any information. It is often acceptable to lose some information in order to achieve greater compression and this proves to be one of several areas of compromise in visual computing. Applications increasingly offer a compression option (such as LZ compression for TIFF files) at the time of saving a file but beware of using a compression type which the receiving application cannot understand. Compression will increasingly be built into hardware and new chip sets support JPEG, MPEG and H261 standards (described next).

As well as using compression within your own system, it can be vital when communicating data to other systems on a network or down a phone line. POTS (Plain Old Telephone System) has been considered to lack the bandwidth needed to handle the demands of intensive transmission (such as good quality video) and alternatives, such as ISDN, are often needed. Compression systems are already in use which will send 320x240 pixel colour images at 15 fps over POTS.

7.6.1 JPEG

The JPEG (Joint Photographic Experts Group) algorithm for still images can compress an image by 25 to 1 with minimal loss of quality, but can take 15 minutes to compress a 25MB image in software (on a 25-MHz 68030 machine). Built into hardware, however, specialised compression processors have the performance to sustain video rates. The algorithms must, of course, be able to decompress as efficiently as they compress and the JPEG algorithm is an example of a symmetrical algorithm which uses the same number of operations for both processes and hence the same time. It is becoming increasingly available as a hardware add-on, and much higher compression rates are possible if some loss of information (often imperceptible) is acceptable.

7.6.2 MPEG

The MPEG (Motion Picture Experts Group) algorithm for motion picture images has been slower to arrive than JPEG but is now being seen in add-on boards. Because of its delay in arrival, a 'moving'

JPEG has evolved in the meantime. In order to take video in to the computer in real time, to store it in an acceptably limited amount of space, and to export it in real time, a lot of efficient compression is required. Interestingly, and perhaps surprisingly, whilst MPEG is designed for use in areas such as video, it is not suitable for non-linear digital editing, which is at the heart of the promised desktop video revolution. This is because the temporal compression compresses moving image data by storing only the difference between one frame and its preceding frame, instead of storing all of each frame. The information from the previous frame is therefore required in order to construct the current frame, and this can only be achieved through linear, rather than random, access.

7.6.3 H.261

H.261 (also known as Px64) is a specification for a method of sending compressed video over digital phone lines and local area networks, particularly addressing the needs of video conferencing. It is likely to become significant since it has been adopted by a number of major companies such as AT&T, Motorola and British Telecom.

7.6.4 IFS

IFSs (Iterated Function Systems) are used to derive a simple set of fractal rules from complex data, such as a picture, and thus allow potentially extreme data compression. They have the advantage over other compression systems of being resolution independent. Whilst the principles have been talked about for several years and research papers published, applications are only just starting to appear in the high street and their impact has yet to be assessed.

7.7 CUSTOMISATION

Any off-the-shelf computer can make some attempt at modelling and rendering, subject to suitable software being available. We have, however, established that our discipline has special requirements, and it is worth summarising the ways in which we might optimise a machine for our particular task.

It rapidly becomes tedious to have a long delay before a model is redrawn after you have moved or amended it and then to see that delay increasing as the model becomes more complex. If the model is in a shaded mode the delay is longer and watching an image being ray traced can be like watching your life tick away. It is therefore desirable to have a fast machine, ideally with a bank of accelerators added.

24-bit colour is great for photographic rendering, but unnecessary for wireframe engineering drawing. A large monitor does, however, make life a lot easier, and I find my 16-inch Apple monitor a very satisfactory compromise. In order to run it in full colour you need to add extra video RAM or buy a suitable video card, depending on the machine that is running it. If I was working on large, complex models I would want a 21-inch+monitor.

You can't have too much RAM installed but for anything at all serious you'll need 16Mb. 32Mb is a comfortable starting point, and it is possible to manage on less, but heavyweight colour graphics can require 128Mb or more for comfort. This is not to say that 12Mb+ virtual memory is not practical for our area, it is just less convenient and not enough to get commercially serious. Bear in mind that even basic modelling software now recommends a minimum of 16Mb and that future packages are likely to want more.

I can't imagine starting without a hard drive, the question is only: 'how big ?' Despite reading that 'size isn't everything', you will find that 60Mb gets full very fast and 2Gb would be nice. I managed with 250Mb but off-loaded files to an optical disc as they built up, and I'm not building models commercially. I also added a second 1.2Gb drive when prices plumetted. Large hard drives do not encourage economical housekeeping but are very convenient. An ideal would be to have two, large, mirrored hard drives so that all data was automatically backed up, and it would take a catastrophic accident to lose it all. I would also employ someone to back up every day to a secondary medium, since I'm no better at doing it than anyone else. A number of applications provide the option for automatic backup and DAT drives still provide relatively cheap storage for gigabytes of data but are being overtaken by magneto-optical and CD. My 128Mb MO drive proves ideal for archiving, but I fail to take my own advice and still store the discs beside the computer instead of in another room (or town!) as I should for optimum security.

I use a mouse for drawn input but strongly recommend a graphics tablet if the budget will stretch to one. The degree to which you need a tablet will be determined by the sort of work you are doing (and also by the background you bring to the work) but the mouse is ergonomically very bad for freehand drawing. Start with a mouse and then discover through doing it how much use you make of it, how much you use numerical input from the keyboard, and how much use you could make of a puck to input data.

You will need to get information out of the computer, but only you can decide whether that needs to be printed, plotted, photographed, filmed or video taped. Be warned that a video set-up can easily cost more than the computer equipment, though this is easier to bear if it is shared by a number of operators.

Although I'd like to claim you can model and render to professional standards on the cheapest, simplest PC, I'm afraid it's just not true. You can get a flavour of the discipline, get to understand its principles and do worthwhile things on simple hardware, but you will need more power to produce the images shown on the covers of the manuals.

7.8 EXPENSE

The cost of achieving a modelling solution can be measured in both time and money, time itself having a price tag in any commercial situation. Direct cost comparisons between platforms are only possible when looking at identical functionality since, for example, the built-in networking ability of the Macintosh will offer no cost saving if the machine is not to be networked. Similarly, the extra power of an SGI workstation is irrelevant if no demanding work is intended. It is also worth taking into account the cost of installation, maintenance and training over a three-year period when costing a machine.

7.8.1 TIME

If you are quietly investigating modelling as a spare-time interest
and are happy to leave your machine running over the weekend to
generate a single high resolution, full-colour image, then the
minimum hardware capable of doing the job is adequate for your
needs. Indeed, there is an added preciousness about a picture that
has been nursed along over several days which is missing from one
created in minutes. It also allows time for you to think about what
you are doing. This might seem too obvious to be worth
mentioning, but when technology speeds up production it also cuts
back on the gestation period which is part of the creative process.
The result can almost precede any clear intention.

In a production environment, however, time is money, and fast
equipment means (potentially) more jobs can be undertaken. Clients
are likely to want a quick turn around, with their expectations being
partly fed by the technology itself. Acquiring a number of machines
and networking them together is often part of the solution, especially
in a business context, and I am constantly aware of the power I am
missing out on by having a single machine isolated at home.

It is also important to have software which will do the job
required efficiently and also to be thoroughly familiar with it.
Although much of the fun of modelling is in solving new problems,
it is much quicker when the problem is a familiar one and the
solution has been rehearsed.

7.8.2 MONEY

Many computer users spend their lives dreaming of the great things
they would do if only they had a bigger/faster machine. It is,
however, very satisfying to achieve the maximum from a limited
machine and this can often lead to greater ingenuity and invention.
Also, with a basic machine, one can be quite confident that projects
could be carried forward onto more powerful machines if the
opportunity arose, or, that connection to a network would give
access to extra power. As a home user the best advice is probably to
decide on the software that you need to run and then to find the

machine that will run it 20% faster than the speed you genuinely require. This will give a little room for leeway but not stretch finances further than necessary. If you are confident that you are going to need to develop your equipment in the future, then you must also ensure that your chosen machine has enough expansion slots. Finally, you can expect to be able to sell or trade-in your machine in order to make another purchase, though you must not delude yourself about the speed at which computer hardware devalues. Too many people wait on the sidelines of computing, waiting for the definitive machine to arrive; whatever you buy will be superseded but that does not make it out of date if it still does the job you require.

If you use a computer professionally, then there are good reasons for spending more money on your system. Hardware is usually going to cost less in the long term than your labour or that of your staff, and there are often incentives to be close to the leading edge of technology. It is well worth talking to your accountant about leasing, since this not only spreads the cost and can have tax advantages, but can give you a high level of flexibility in upgrading equipment when you need to do so. In education, leasing is increasingly considered now as a means of establishing an immediate hardware base, rather than having to build it up over a period of years (and thus losing out on the use that could have been made of it in the early years). If you are relying on a PC in your business, it also pays to consider a maintenance contract. Since equipment is increasingly reliable, it is difficult to bring yourself to spend that extra five per cent at the outset, but you have to assess the cost to your business of being without equipment for the period of any repair.

I know of a number of businessmen who have been talked into spending far more money than they needed to on their computer systems. They have ended up with very nice systems but specified far beyond their real requirements. I do my word processing on a once powerful computer because that is my machine, but if I only did word processing I would trade it in straight away for the cheapest available – I would be wasting its power. Unless you have unlimited funds available, there is always a compromise to be made between the ideal system and the one you can afford.

7.9 ERGONOMICS

At the same time as considering hardware, it is worth thinking about how and where it is going to be physically used. In the broadest sense, the working environment must be psychologically sound as well as ergonomically effective. It is not within the power of the system's designer to determine whether the operator works in monastic silence or with Wagner playing at full volume, nor whether the lighting is from soft uplighters or glaring spotlights, but these factors form part of the total user interface. It is also the case that scientific disciplines are likely to be carried out in a very different environment from that of the creative world and that the personnel will have very different backgrounds. Someone with a scientific background, who has probably become familiar with computers and programming as a general purpose tool, is likely to be more comfortable inputting numbers through a keyboard than a graphic designer with, perhaps, no background in mathematics. The designer, on the other hand, will feel immediately comfortable holding a stylus that may be alien to the scientist, yet they may both need to use the same computer system.

When a process such as drawing a freehand circle, with which we are all familiar using pencil and paper, is being undertaken on the machine, it is easiest to use a tool that simulates the process with which we are familiar. In that instance using a stylus is manually identical to using a pencil (but with the initially disconcerting difference that you don't see the result of your drawing on the pad, but on the screen in front of you). A fresh coordination needs to be acquired between hand and eye. A mouse also requires the hand to be moved in a circle, but, being gripped differently, excludes the subtle finger control we would normally expect to exercise. Cursor keys present a further level of removal from the real-world experience and sitting down to write a program to draw the required circle is probably as far removed as you can get from using a pencil.

It is not, therefore, possible to describe an ideal environment, as it depends on who is doing what with the system. Certain features can be recommended, however, not just to make the job comfortable, but also to get the most out of the user. The following should be set according to recognised ergonomic standards:

1. The height of the desk and chair must be set relative to the operator when resting the balls of the feet on the ground, and not arbitrarily. The chair seat should be able to tilt forward.

2. The position of the keyboard, pad, mouse, etc., relative to the operator. RSI (Repetitive Stress Injury) is the subject of medical discussion and keyboards are claimed to be a major problem. Long periods drawing with a mouse can also be a problem (as I have discovered).

3. The size, distance and angle of the monitor.

4. The radiation level of the monitor (new monitors conform to high standards, old ones can often be converted).

5. The monitor's resolution and refresh rate (72 dpi is a standard resolution, but higher is good for fine work; higher refresh rates mean less flicker).

6. The ambient and local light level and type (low light levels allow the monitor brightness to be turned down, thus relieving eye strain, but might conflict with other office needs). Strip lighting can flicker, especially when nearing the end of its life, and seems to be a particular problem when combined with monitor flicker.

7. The continuous time spent at a monitor: various recommendations exist and include the advice to get up and move around during breaks.

8. The combined noise of collected computer and peripheral fans should be recognised.

There are now specific EEC standards which must be met and stricter standards upheld elsewhere, notably in Sweden.

CHAPTER 8

COMMUNICATING WITH THE COMPUTER

It becomes so obvious to tap at a computer keyboard, wiggle a mouse and watch the results of these actions on the computer screen that it is easy to take for granted that this is the way one 'talks' to a computer. It is, however, worth considering the broader view – that these are merely examples of the input and output of data. However sophisticated the data may seem to us, perhaps a photographically realistic three-dimensional scene, the computer can only distinguish between 0 and 1. It is the job of the input and output devices to work with the computer and create an interface that makes our communication with the machine as friendly and efficient as possible.

8.1 INPUT DEVICES

An input device is a piece of hardware which allows us to put data into the system. This input may be coded (such as a typed instruction) or positional (such as traced off a map). As modellers we might often find ourselves concerned with inputting data both in drawn and numerical form. My first experience of computers (many years ago) was of hours spent punching cards in a noisy, communal punch room, followed by the presentation of several boxes of cards to the computer operators through a mysterious hatch in the wall and a day or two waiting for a print-out to be returned.

If one hole was punched in the wrong place on one card then the program didn't run. Not the environment for which most designers would choose to abandon their drawing boards and putty rubbers.

8.1.1 KEYBOARD

The ubiquitous keyboard is so familiar that it is easy to forget how it works. It usually conforms to the same layout of letters, figures and symbols as the typewriter (with some regional variations amongst the non-alpha-numeric characters). The QWERTY keyboard (named after its first five keys) is the standard; ergonomically superior key layouts have been designed but retraining presents too great a problem for them to be implemented in most cases. Keyboards are also available which improve on the standard rectangular format and help to avoid the stress injuries which can arise from constant use. It is possible, with a non-manual keyboard, for alternative key functions to be held in software and switched between. Your keyboard Control Panel, for instance, will enable you to choose between a UK and a US keyboard, if both layouts are installed.

A number of keyboards are available for PCs, some with an additional numeric keypad to speed up numerical input, four cursor keys which move the screen cursor left, right, up and down; and 'function' keys which can be configured by the user (or the current application) to do prescribed jobs. Depression of a key, either on its own or in conjunction with another (e.g. SHIFT, CONTROL, ALTERNATE, COMMAND) generates a unique electronic digital code which is interpreted by the computer's CPU. The code is normally the international standard ASCII (pronounced 'askey').

If a key is held down too long it will send its message more than onceeeeeeee (this is because the keyboard is checked by the CPU about 50 times a second to see if any key has been pressed). Whilst the system may not respond to the repetition of messages from some keys, it is likely to do so with all the alpha-numeric keys, which can give rise to errors. Once a key has been recognised as having been pressed, the significance of the code is considered according to an established priority, so that QUIT or BREAK, for example, may be given priority over everything else that is going on. An interrupt which is given priority over everything else is called an NMI (Non-Maskable Interrupt). The key 'repeat rate' and 'delay until repeat' can normally be set by the user.

8.1.2 MOUSE (+ variations)

The mouse is another ubiquitous device that fits in the palm of the hand and is rolled over a smooth surface. In a mouse with a mechanical mechanism (the most common type), a ball in the base of the mouse is rotated by the movement across a surface and these rotations are translated into data which moves a screen cursor on a corresponding path. The less common optical mouse establishes its position by detecting reflections from a beam shone downwards by a light-emitting diode, and a reflective surface is therefore required. Mechanical mice collect dirt that interferes with their efficiency and should be periodically cleaned; they work best on a clean mouse mat.

Move the mouse to the right and the screen cursor moves a proportional distance to the right, move it forward (assuming a horizontal surface) and the cursor moves up the screen (assuming a vertical screen). The screen cursor, incidentally, is often not the same as the keyboard cursor and may change its form according to the function it is currently fulfilling. The speed of tracking of a mouse, and its double-clicking speed, can be customised in a Control Panel to suit the operator. The PC mouse may have two or three buttons which will be detailed to undertake different functions according to the application in use. The mouse is connected (often to the keyboard) by a flexible lead which can be altered for right or left-handed operators. Cordless mice are now available, which are less restricting, more expensive and ambidextrous. In general the mouse can be regarded as a cheap, simple and relatively low resolution device. It is important to remember that the coordinates returned by the mouse are relative to the position on the last occasion that the ball was turned. If the mouse is taken off the surface its absolute position is lost.

The 'tracker ball' is similar to an inverted mouse and is operated by turning the ball with the fingertips. It moves a screen cursor in the same way as a mouse but is, perhaps, less immediately intuitive to use. One of its advantages, however, is that it remains stationary and for this reason has become incorporated into some laptop computers. As the size of the ball is increased, so subtle movements become easier whilst big movements require the ball to be spun more. A giant tracker ball has been suggested as a control device for handicapped users who have difficulty making fine movements.

8.1.3 2-D GRAPHIC TABLET

Graphics tablets have a pad, normally between A4 and A0 size, on which the position of a hand-held sighting device called a 'puck', or a pen-like 'stylus', can be detected with great accuracy. They can also be used with digitising tables of much greater size for higher levels of precision. The puck has a small window showing cross hairs, which are to be aligned with the current data point, and buttons to determine the use that is to be made of that data. Either the pad or the puck transmits continuous signals which the other receives, and can be electromagnetic, electrostatic, ultrasonic or infrared. The signals translate into X,Y positions with an accuracy of up to 0.001 of an inch and the pads are widely used as a means of transferring data from a drawing on paper to the computer. Before starting work the digitiser must be orientated to the sheet of paper, so that verticals remain vertical and the origin is correctly located.

If the puck is replaced by a stylus then the same principles allow freehand drawing, which might prove useful in a paint program. Whilst the puck is often used to input data without reference to a display monitor, the stylus operator normally uses the monitor for positional feedback. The use of the stylus is associated more with continuous movement and the puck with the input of discrete points. Some styluses are pressure sensitive, which is a great advantage in trying to match the subtlety of normal hand media, but this facility is dependent on the software's ability to support it. Both puck and stylus are normally connected to the pad with a flexible lead, though cordless versions are now available with the disadvantage that they are easier to mislay.

The digitising pad can often be configured by the user, allowing the relationship between its drawing area and the screen area it maps to be flexible. Some applications cover part of the pad with menu overlays relating to program functions, and when the stylus or puck is used to select from this menu it is said to be used as a 'pick'. The same applies to the use of pick or stylus to move the screen cursor in order to select from any screen menu displayed. The stylus, in particular, is often used to select data from the current screen image for further treatment, for example in a paint program to select one colour from the screen image for use elsewhere.

8.1.4 3-D DIGITISER

The systems described so far have been used for inputting 2-D information and this may, of course, provide the basis for building a 3-D model. It is, however, possible to input 3-D information directly, though the equipment for doing so is not yet widely found. The position of a stylus tip, moving in 3-D space, can be tracked using acoustic, mechanical, optical or electromagnetic means. An object can thus be traced over with the stylus, but care must be taken to select logically suitable points on its surface for digitising. A new device for the input of three-dimensional information into a computer is a sonic digitiser using four microphone sensors to calculate distance from a stylus tip emitting an ultrasonic signal at a rate of up to sixty times per second. Commonly, the object would have a mesh grid drawn over its surface and the intersection points would be those digitised. The process is, as yet, time consuming and painstaking. Laser scanning provides an alternative method, but all current systems have problems with objects containing inaccessible areas.

8.1.5 TOUCH SCREEN

There are occasions when it is desirable to have the most direct possible relationship between alternatives open to the operator and the means of attaining those alternatives. Such a high level of interactivity is appropriate, for instance, if the public has access to an unattended machine, which they are to be encouraged to use intuitively to get information. A touch screen allows the user to press the screen and get the same response as a mouse-click would normally give; buttons simulated on screen can therefore be 'pressed', and items 'dragged' (if so programmed) with a fingertip. There are a number of different technologies that can be used to build touch screens and the choice is likely to be based on cost, sensitivity, robustness and available size. Monitors can be supplied with touch screens built in or they can be added subsequently.

Whilst there is no frequent case when this is of specific use to modellers, it is increasingly the case that computer models may subsequently be made accessible to the public through such interactive media.

8.1.6 SCANNER

We are likely to need to import complete images intact from a range of sources as well as to create them by hand (e.g. to use as image maps). Flat artwork can be imported using a 'scanner', in which the material is placed face down on a glass plate and scanned using similar technology to a photocopier, producing a file in a common format. Until recently, colour scanners were very expensive but the price has dropped considerably and a 24-bit colour scanner able to work at 600dpi (dots per inch) can be bought very reasonably. Ultra-high resolution scanners are likely to be found only in bureaux, but desktop machines are coming out with 2400dpi although it is worth remembering that both this resolution is usually by software interpolation rather than an optical resolution, and that the highest resolution might not be possible when the whole flatbed area is scanned. Scanning at that resolution can also create massive files.

The desktop machine's usual image size is about A4, and small hand scanners are available which cover a width of about four inches, whilst being dragged manually down the image. Problems in using them can arise if the copy is slightly misaligned or, in the case of the hand scanner, the dragging is not even, but the latest software improvements have largely overcome these faults. When text is imported, the use of OCR (see below) software can produce a portable ASCII file (allowing, for example, a newspaper page to be scanned in and converted to a standard ASCII file which could then be reset using a typographic package). Machines are available on the market which make use of the shared technology of colour scanners, printers and photocopiers to do all three jobs with consequent cost saving but with potential loss of flexibility in a shared environment.

8.1.7 NEW INPUT DEVICES

Three forms of input which are developing fast are OCR (Optical Character Recognition), in which an intelligent system reads printed (or sometimes even hand-written) text; direct speech, in which a system understands human speech (in a currently limited vocabulary); and the DataGlove, which translates movements of the operator's hand (encased in the DataGlove) into comparable movements in the three-dimensional space of a robotic hand which might exist or merely be a computer model. The DataGlove is one tool of virtual reality in which the operator is able to enter into a three-dimensional computer-generated space. (The potential and implications of virtual reality are so great that it will be dealt with later in the book.) A directional hand tool (a palm-held ball shape, tapering to a pointer, with input buttons) is one of the devices which has been developed to allow someone in a virtual environment to 'fly' through the virtual space that surrounds him.

8.1.8 VIDEO

It is common to import material from a video source, which might be either direct through a video camera, or using pre-recorded material through a VCR (Video Cassette Recorder). The suitability of this as an image input medium depends on the use to which the image is to be put, since broadcast video resolution can be fine for producing material for TV viewing but not necessarily for reproduction as a print. The quality of the image is determined by its spatial resolution (768x576 for broadcast TV in the UK) and by its colour depth (i.e. 24-bit represents 'full' colour). A specialist video card is needed to enable the computer to interface with a video input but these are now starting to be built into new machines, especially those intended for multimedia work. A broadcast quality card can cost more than the computer itself but good quality cards of lower specification are now reasonably priced.

Live action can also be grabbed-in, the main problem being the massive file sizes taken by large screen, full colour video at 25 frames per second in the UK using the PAL system (or 30 fps with the NTSC system in the USA and some other countries). These problems are now being alleviated by hardware compression boards capable of taking in live video and storing it with up to 160:1 compression. Computer-based video is an area of great recent progress and the market's interest in multimedia is hastening this development.

8.1.8.1 STILL CAMERA

Increasingly available are still cameras designed to capture static images on a digital medium, using either discs or RAM, at a potentially high resolution. If the material shot is exclusively destined for use in the computer then this is a fast, reusable medium, though still relatively expensive for professional standard resolutions. There is much talk about digital photography superceding traditional paper-and-silver methods, and it seems certain to do so in many areas. Being able to grab images of the real world directly into the computer, manipulate them digitally and integrate them seamlessly with other media, suggests a very productive environment.

8.1.8.2 LIVE CAMERA

A vertically mounted video camera is usually to be found in the corner of computer graphics studios standing ready to import flat art work, usually being a professional 'three-chip' camera. It has the advantage over a scanner that it can also be used for three-dimensional subjects and it is easier to view the image on screen while it is being manipulated, but is not necessarily of a suitably high resolution. A domestic camcorder can equally well be used, though at a slightly inferior quality. For preliminary visualisation work, and often for work destined for domestic video, the simplest equipment is adequate and easy to use. A camera can also be used for grabbing-in live action if a suitable video board is installed.

8.1.8.3 VTR

Pre-recorded material can be input from a video tape deck; either a separate unit or one integral to a camcorder. The same issues of quality apply. If it is required to input video frame by frame, perhaps so that each frame can be treated in a paint system, then professional equipment and a video controller are needed. These are likely to be far more expensive than your average PC, though demand is pushing prices down.

8.2 OUTPUT DEVICES

Being able to input data only allows for a monologue and in order to communicate with the machine it must be able to 'talk back'. It does this via a possible range of output devices. A printer or plotter can provide 'hard copy' on paper; film and video cameras can record still or moving images; or real time feedback can be given by the monitor screen. Output can also be of data to a storage device but devices such as disc drives have been dealt with in the previous chapter.

8.2.1 VDU

The basic VDU (Visual Display Unit) is normally a standard computer monitor (although a video card might also display its output on a video monitor or television). Its screen is such an integral part of the way we work with computers that it is often forgotten that it is an output device and not a window on the computer's mind. It provides a quick and understandable, but not always accurate, view of the state of progress within an application.

A normal monitor cannot, for instance, show a straight diagonal line without the 'staircasing' effect of a pixel screen, and an image of photographic quality can only be shown at the much lower resolution of the screen itself.

A number of properties of a monitor are worth considering, particularly size, resolution and colour. The resolution is usually expressed in dpi which indicates how many pixels there are to an inch; the more there are, the smaller they obviously are, and the clearer the picture definition. Typically a modern screen will have at least 72 dpi, which used to be described as high resolution, though 80 dpi is not uncommon. Note that it is the dpi figure (or increasingly the 'dot pitch') which indicates the screen's resolution and not the pixel count, which is often quoted instead. A 20-inch screen with 1024x768 pixels is the same resolution as a 14-inch screen with 640x480 pixels, and of a lower resolution than a 15-inch screen with 640x870 pixels (72 dpi compared with 80 dpi). You will notice that the first two sizes mentioned are of a horizontal format compared with the vertical format of the third monitor. Horizontal is the standard, but for DTP it is convenient for the screen format to match that of the layout. This 15-inch monitor is, therefore, designed with a vertical A4 sheet in mind, whilst the 20-inch monitor permits the viewing of a double A4 layout. Screens which rotate from vertical to horizontal give you some advantages of both formats. Portable machines have different screen technologies, and although they are improving fast, cannot yet match the clarity of conventional screens.

A large monitor can make working life much more comfortable, but the size actually needed is best defined first in terms of the job to be done and then translated into inches; bear in mind that monitors are generally described by the diagonal measurement of the tube, which will be greater than the available viewing area. If your work involves a lot of fine detail then the highest resolution helps, though the zoom factor in all applications can be used as an aid. (The viewing distance of the monitor is also a factor.) The equipment you buy will usually be chosen by compromising several factors. The other major factor with monitors is colour.

The most basic screen will display only black and white and simulates greys with graded black and white patterns. A greyscale

screen can generate a range of greys but no colours. An 8-bit screen
can produce 256 colours and a 24-bit screen generates more than 16
million colours (although you won't have enough pixels to use them
all at once). For close to photographic quality a 24-bit display is
needed, but the results at 12-bits (4096 colours) are quite good and
can be very good if the palette is judiciously chosen. The difference
in palette size is most easily seen in areas of gentle gradation, where
small palettes give rise to banding. It can be a waste of time and
space to generate images with larger palettes than can be displayed
on the monitor for which it is eventually destined (is it worth ray
tracing at 24-bits for your monitor if the image will end up in an 8-bit
application?). For precise colour work it is possible to accurately
calibrate the colour seen on the monitor to the colour produced by
another output device, such as a printer.

Unfortunately you cannot just plug any monitor into any
computer and start work, the machine must be capable of driving the
display. There are limits to the size/bit-level of monitor that any
computer can drive with its basic capacity and it may be necessary to
add video RAM (VRAM) or a separate video card in order to drive
the monitor of your choice. A phenomenon that can seem strange at
first is to see two monitors being driven by one computer, so that the
cursor passes from one screen to the other during horizontal
movements. This can be useful, for example, to display a number of
application menus and control windows on one screen whilst having
the main image window on a second screen, therefore avoiding a
single cluttered screen.

Video work requires further consideration, and if you anticipate
that this will be a requirement, it is as well to check out your
proposed purchases thoroughly. A computer monitor is non-
interlaced, which means that every line on the screen is scanned at
each pass, whereas a TV screen is interlaced, meaning that only
alternate lines are scanned on each pass. The most noticeable results
are that our computer monitor is comfortable on the eyes, having no
perceptible flicker, and that a TV image will not display thin
horizontal lines without dramatic flicker. This requires
consideration when producing images for video, as static, single
pixel horizontals must be avoided . For example, a computer
window can look terrible on a TV if the window 'frame' is made up

of single pixel lines which flicker badly, although the contents of the window might look fine. Anti-flicker filters are fitted to some video boards to alleviate this problem and are a great improvement for general use, but the resultant signal is no longer strictly of broadcast standard. A multisynch monitor (which can adjust to different refresh rates) is often required since it, and it is worth noting that for video work a resolution of 576x768 pixels matches a TV screen. A computer can also generate colours outside of the range that a TV can handle and rich saturated colours are usually to be avoided. It is possible to run images through a PAL filter (as in the UK) or NTSC filter (as in the USA) which will strip out 'illegal' colours, but since TVs are often poorly tuned, it can prove safer to err on the side of conservatism rather than risk glowing reds. The net result of these differences is that images generated only for the computer often look terrible when transferred to video.

Since you are likely to spend a lot of time working at your VDU it is an important purchase, but the technology is so improved that there are many fine monitors available. I'm delighted with my 17-inch monitor and was surprised to find that it did not receive great reviews in comparative tests, but this indicates that there is a high subjective element in screen choice; it pays to have seen and used your screen before purchase. Other considerations in making your life comfortable are a high refresh rate to minimise flicker (75Hz is common and satisfactory), anti-glare coating (particularly in a bright environment), low radiation levels (modern monitors conform to stricter guidelines), and a tilt and swivel base (almost standard and ergonomically essential).

8.2.2 PRINTER

An obvious output device is the printer, of which there are several types. The cheapest and most popular is the 'dot matrix' printer, which strikes the paper through an inked ribbon with a number of fine pins. The pins are carried in a moving head and the image is made up of a number of small dots, the resolution of which is determined by the number of pins in the head and the subtlety of the software driving it. A basic printer has 9 pins, 24 pin machines are now common, and higher pin numbers have recently become available. Many dot matrix printers can be equipped with colour ribbons (comprising bands of magenta, cyan, yellow and black) through which the pins selectively strike, but the results are not very convincing and are liable to be inconsistent as the ribbon fades. The dropping price of other technologies is causing the dot matrix printer to be superseded especially for non-text applications but it provides a cheap means of getting a rough print out.

Laser printers employ electrophotographic technology (first developed in photocopiers). They are more expensive but produce very good quality results in black and white, until recently with a typical resolution of 300 dots per inch (dpi) but increasingly with 600dpi as standard. They are also available with higher resolutions and with colour – at a price. The human eye can distinguish separate dots up to about 1000 dpi. Ink jet machines (which shoot precise jets of coloured ink onto the paper) and bubble jet machines (which burst bubbles of ink onto the paper surface) are reasonably priced, quiet, and give clean colour, whilst thermal transfer printers (which use special paper) are effective but relatively expensive. Dye sublimation printers are becoming affordable and produce near photographic results but are more likely to be used via a bureau. It will always be the case (with foreseeable technology) that screen colour will fail to match printed output accurately, owing to the different ways in which the images are formed, but calibration brings them closer.

The means of creating colours on the screen is fundamentally different from that used to create printed colour. Screens optically mix from the three 'additive' primaries of Red, Green and Blue (RGB) to make all colours, the three primaries combining to produce white. In the 'subtractive' system the three primaries of Cyan, Magenta and Yellow (CMY) are overlaid to make any colour, the three primaries combining (theoretically) to produce black. In practice the print process requires the addition of black (K) pigment and the system is referred to as CMYK colour. Since the RGB screen image will often be representing a CMYK printed image, accurate matching of the two systems is vital though not always easily achieved (and since the colour gamut of the two systems is different a perfect match is not always technically possible). Pantone has produced a set of matches between printing colours and screen colour which is helpful in areas such as DTP, but a precise match requires particular combinations of screen and video card under controlled conditions.

8.2.3 PLOTTER

Plotters produce hard copy by drawing on paper with pens which can move along the X axis and be raised from, or lowered to, the surface of the paper. Either the pens can move along the Y axis or the paper can be moved along that axis past the pens. 'Flatbed' plotters hold the paper flat in either horizontal or vertical plane and can vary from A4 size to 10 feet long. 'Drum' plotters take less floor space than horizontal flatbed plotters to produce large plots on cut or continuous roll paper, by wrapping the paper over a drum roller, or by moving it from one roller to another. The plotter usually has between four and ten pens, selected under software control. Since the pen can produce only lines and dots, and must combine these styles to produce shading, it is best suited to linear or diagrammatic images only. The software that drives the plotter cannot convert raster images to plotable form. The best plotters are very high resolution devices, widely used in industry.

8.2.4 VTR

Video tape is the most commonly used medium for recording moving images today. A large number of formats are already available and more are developing, providing great increases in quality. Whilst it is still generally true that the larger the tape width the greater the potential quality, improvements in tape technology mean that excellent results are now practical on narrower tapes which are associated with smaller (often cheaper) machines, though not necessarily to broadcast standard. Traditionally, tape has been used for analogue recording, but increasingly digital technology is invading the market place and can be expected to grow in influence over the next few years. Digital storage allows images to be subject to manipulation without loss of quality and is in the form needed to be handled by computers. Digital video equipment now has a foothold on the market.

There are many video formats which reflect the range of qualities, sizes and prices of systems available for use from home to professional broadcast standard. A significant distinction is between those that store the red, green and blue signals separately (component) and those that merge them together (composite). The highest standard professional formats are component but broadcast transmissions are currently composite.

Assuming that video is being used to record moving images, the ability of the computer to generate them (or store them and then replay them) in real-time is crucial. Although mentioned already, it is worth noting that the computer cannot always calculate and display subsequent images at 25 (or 30) fps and if that is the case then the frames must be 'dropped' individually to the VTR. This capability requires a professional machine and also an animation controller, which can be either in software or hardware. Both are likely to be expensive.

8.2.5 STILL CAMERA

Setting up a still camera and photographing from the monitor screen is a cheap and simple way of saving images. The camera must be set up on a tripod, using daylight balanced film, a shutter speed slower than the screen refresh rate, and ambient light must be prevented from falling on the screen and causing reflections (usually by improvising a hood to enclose screen and camera). It is also desirable to run tests to find the ideal exposure and screen contrast to set, the maximum contrast available on the screen being outside the ability of film to record accurately (colour prints having a lower contrast ratio than transparencies). The maximum resolution will be that of the screen, however, and the scan lines may be more obvious in the photograph than on the screen itself. There is also likely to be some distortion of the image owing to the curvature of the screen. An old, but effective, technique was to mount the monitor on a rubber mat and to bang it during the exposure!

In order to guarantee the highest possible quality images, special film recorders have been developed. These contain a very high resolution black and white monitor with a flat screen and three coloured filters, to which the camera must be accurately aligned. The three colour planes in the frame buffer are separately displayed through the appropriate filter and recorded onto the same piece of film. The resultant resolution is far greater than on a colour monitor since the black and white screen is not restricted by the shadow mask required by the former. Bureaux are available to produce slides from your own files at a resolution of 4000 to 8000 lines.

8.2.6 FILM CAMERA

The same principles apply to recording sequential images on cine film, with the added problem that the camera must be capable of reliable single framing, preferably under the control of the computer. It is not practical to film moving images of any quality from screen in real time because synchronisation problems between the shutter and scanning are added to those described for still images. Recording directly from screen with a video camera normally produces poor images, although some professional cameras allow the shutter to be adjusted to the screen refresh rate for optimum synchronisation.

8.3 THE SCREEN INTERFACE

The role played by the screen interface has been touched on when describing the operating system and the specific problems of a modelling environment, first mentioned in Chapter 2. Its importance cannot be understated, and in the chapter dealing with specific applications I will often refer to the way in which the programs allows you to interact with them. The ease with which you are able move around and between the different features of the application and to set properties such as position, colour and surface, contributes to your enthusiasm for the job at hand and to your level of productivity. One particularly interesting problem in a modeller is the need to navigate 3-D space (and 4-D space, if you include the temporal dimension of an animation) by reference to a 2-D screen.

The need to position objects in space by rotating them around X-, Y- and Z-axes has found different solutions. Sometimes the three axes are controlled similarly but separately, perhaps by knobs or by the mouse after the appropriate axis icon is selected; but more often rotation about the axis orthogonal to the screen plane (i.e. the Z-axis) has a different control. Rotations around the X- and Y-axes are typically controlled by virtual sliders aligned along the X- and Y-axes, or by horizontal and vertical movements of the mouse (because horizontal movements to left and right can be seen as analogous to turning a freely held object around the Y-axis and vertical movements around the X- axis). It is less easy to find a method of mimicking rotation around the Z-axis which is consistent with those of X and Y, and a less intuitive use of the same devices is often implemented.

Research on the problem has produced a number of inventive solutions which represent the movements in terms of circles or spheres and different manufacturers have their own preferences. The movement of the object might be made in real time, in which case control of positioning can be interactive and it is likely that numerical positioning will be available for greater accuracy. The axes around which the object rotates by default are normally the world axes, which are unlikely to coincide with any of the object's axes (except when the object is in its canonical position or has been shifted without rotation). The pressure sensitive ball device, that has already been mentioned, represents one mechanical solution to the problem.

None of these methods matches our own use of binocular vision to interpret spatial position. Our brain uses the difference in information received from our two eyes to understand depth, and even slight head movements can be enough to clarify spatial ambiguities. Stereoscopic viewing is available for use in conjunction with computers but tends to be used for viewing the final result

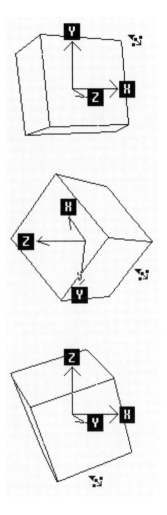

Rotating object shown with labelled axes of rotation

rather than for interactive creation and manipulation. The method used involves replacing the single screen image viewed by both eyes, with separate images for each eye. Spectacles can hold separate mini-screens in front of each eye or can be used to filter separate information for each eye from a single screen by the familiar red/green lenses or by polarising lenses at right angles to one another. Alternatively, the screen information intended for each eye can be presented in turn whilst the lens covering the opposite eye is blacked out electronically, the synchronisation being controlled by the computer presenting the images. If the information for both eyes is shown on a single screen, then the display can be confusing for a viewer without the appropriate spectacles.

Virtual reality tools will allow you to enter into the same space as your objects and manipulate them from there, though this may not always prove preferable to viewing the scene from the outside. One should distinguish here between 'immersive' VR in which the operator enters into the computer created space, and 'non-immersive' VR where the operator observes the space through the window of the screen, but can manipulate objects within it. It is open to discussion whether the latter is a sufficiently new paradigm to be called VR.

Another spatial problem arises in setting the position and direction of light sources in the scene and there have been a number of very different methods employed by various applications. If the light source is not directional then only the position of the light itself has to be established, and methods vary from placement in top/side/front windows to dragging a patch which represents the light around the surface of a sphere which stands for the boundary of the object's world. It is also possible to position the light, not by setting its position in space, but by demonstrating its effect on a symbolic object in the scene, as seen from the current viewpoint.

A directional light has to be similarly positioned in space but also has to have the direction of its beam established, and probably the angle of the beam as well. This is quite effectively done in some applications by siting the light in the 3-D environment and dragging out a line representing the beam direction vector from the light to another point in 3-space. If the beam angle is variable then lines can be used to represent the edges of the beam instead of the beam's

central direction, which has the advantage of increasing the accuracy
with which a beam can be positioned.

CHAPTER 9

METHODOLOGY

Since a computer model never actually builds itself, it is desirable to consider the methodology – the orderly arrangement of ideas – that needs to be brought to the modelling process at all stages.

9.1 PLANNING

There are usually at least half a dozen different ways of creating a model within any one modeller. Even a B-rep cube, for example, can be created by extruding a square template, by lathing a four-sided object from a half-a-square template, by skinning, by lofting or by selection from a library of primitives, to name just a few obvious methods. The most appropriate method to use in building an object will depend mainly on the facilities and efficiency offered by your modeller and on the use that is to be made of the object.

If it is intended to 'walk-through' the finished model it will probably be advantageous to create it in its simplest practical form (i.e. with walls having no thickness and doors having no moulding) so that it can be redrawn fast enough to achieve a sense of movement through the space. If the final model is to be rendered as a high resolution colour slide, then detail might be more important than file size and a long rendering time acceptable. If the model is to be exported to another package then it is necessary to check that it will lose no important properties in the process of saving to the chosen file type. If a number of separate elements go to make up an object they might be 'locked' together, but will the individual elements be able to be 'unlocked' if a change is required at a later stage?

All these possibilities, and many more, are obvious when pointed out, but can easily be overlooked if the project is not planned. Whilst there is probably no 'wrong' method of working, there are certainly methods which are more productive and less prone to unwelcome surprises than others. It is necessary, therefore, at the outset of the project to have at least an idea of what you aim to do, and, ideally, to have a clear plan of what you are going to do. Experience will contribute much to any decision about working method, particularly when there is a high level of similarity in models that you build and you will establish a routine. Half the fun of modelling, however, is in working out the best method of building a new and unfamiliar object, and, on completion, in realising a better method for 'next time'.

It is also possible that an open-ended and experimental approach will yield fresh discoveries but in real terms a balance is needed between such an approach and that of mindless repetition of a known formula.

9.1.1 'WORLD' MAP

I find it helpful to have a quick sketch alongside me which shows a cube representing the modeller's universe (or relevant part of it if it is infinite) marked with the axes and their rulers. The sketch could, of course, be drawn on the computer and viewed in a separate window, but I seem to be happier with paper. On this I roughly position the object(s), light(s) and camera(s) and with this guide alongside me I rarely get lost in the 3-D space. This might sound like an absurd confession, but in the modellers that I end up using most often it is still easy to get disorientated when moving the camera around and to lose a clear sense of the relationship of the objects to the light sources, for example. The orientation can be regained and/or the relative positions reviewed, within the modeller, but not always conveniently. Some modellers are much better about preserving a sense of position than others but other relationships can be misinterpreted and I wouldn't throw away your notepad just yet.

9.2 MATCHING METHOD TO TOOLS

It is necessary to recognise what your modeller is good at doing, what it is not so good at doing and what it just can't do (though with 3D Studio that isn't likely to be much). You can probably build something very similar to the model you want with any modeller but working to its strengths saves time and heartache. Whilst spinning and extruding are common to most packages, there is a limit to the objects that these processes alone will create, and it is no use being impressed by sample models of wine glasses and office desks if you are going to want to create frogs and curtains.

Manufacturers usually give an indication of the sort of customer at whom their product is aimed, though they are likely to suggest as broad a range as possible. It is usually clear, however, if a product is intended for architects, engineers, visualisers or for less dedicated use. Since you are reading this book you are likely to already own 3D Studio so your decision is made, but this does not mean that you won't be looking out of the corner of your eye at other modellers. When making comparisons between modellers some of the properties you might look for are:

- the ability to create complex curved surfaces (probably with splines)
- the ability to create asymmetrical objects
- the ability to join objects
- the ability to cut objects
- the ability to enact Boolean operations on objects
- the presence of low-level editing tools

If these facilities are not present, you can start thinking of ways to skirt around their absence, probably with varying degrees of success. For instance, if you can't do a Boolean 'addition' operation, can you 'glue' or 'group' or 'link' the relevant objects, and if so, does this alternative give you sufficient functionality? If you can't obviously create asymmetrical objects, can you rethink your object in terms of a collection of symmetrical parts? If you get stuck for a solution then perhaps you have become locked into one way of thinking and need to try and see the problem again simply – as if with the eyes of a child (or move to sections 9.5 and 9.6).

9.3 THE NAMING OF PARTS

Your model may be made up of a hundred separate parts. You might have several versions of your model stored on the hard drive. When you create another part and are prompted to give it a name, resist the urge to call it 'x1', 'thingy' or 'newbit' because, descriptive as the names seem at the time, their meaning will be forgotten by tomorrow, if not sooner. 'LKnee' is more meaningful than 'LK'. Give everything a name which would identify it to a stranger, or else have a clear and consistent coding system for part names (e.g. W=window, 3=third floor, N=north facing, etc. according to your needs). The extra time taken in typing in a clear descriptive name is saved many times over when it comes to locating something in the future (even if the system does not allow us the file name lengths we would like). Using the date as a suffix can be helpful in distinguishing between versions, although the system keeps track of the file's date anyway.

If you need to load two different versions of an object in at the same time, bear in mind that the application might perceive a clash of names if both versions have the parts or objects similarly named. Whilst this can be solved by renaming, it can be tedious to find the only way to proceed is to be presented with every name in sequence and asked for an alternative.

Applications, especially drawing packages, often support 'layers', which can be thought of as transparent overlays to which different parts of a drawing or model can be allocated. These can provide a good way of separating different elements from one another. For example, the floor plans and pipe runs for a building can be held in separate layers and viewed (or operated on) either individually or overlaid on top of one another.

9.4 SALVATION

Everyone knows that you must save your work at regular intervals, that you must make a back-up copy on to a separate medium, and that you must keep a second copy of important files at a different location from the primary copy. Everyone also has his/her own story about losing an hour's or day's or week's or lifetime's work because the cat stood on the delete key or workmen cut through the power cable. I have become good at saving regularly to the hard disc, but always wait longer than I should before making copies on to another medium (despite now having an optical drive to make things far easier). When you lose your work (and you will), you have only yourself to blame.

If you can afford it, a great first line of defence is a pair of mirrored hard drives (described in section 7.2.3), where all data is simultaneously stored on two separate hard drives. There is also plenty of cheap software that will automatically save files at pre-set intervals (such as a simple public domain program) and such software will often perform backups to predefined storage devices. (Take care with software that does a regular 'Save' without presenting a dialogue box because if you have loaded an existing file to use as the basis for a new one the first save operation will overwrite the old file and you therefore need to save it under a new file name before starting to modify the file). Establishing a habit of saving every 15 minutes and backing up at the end of every work session is a good start. In a communal environment your work can be destroyed in so many different and exciting ways that saving it regularly becomes vital, and paranoia on the matter is to be encouraged. A code of practice should be given to anyone joining the community – and enforced.

It is also necessary to consider the lifetime of data held on different media and under different conditions. Data doesn't stay intact forever on most common media and even ensuring a ten-year life span requires thought. Magnetic storage on tape or floppy is not only vulnerable to obvious magnetic influences like the big magnet in your monitor, magnetic fields in your graphics pad (depending on the technology it uses) or to heat sources like the radiator beside your desk; data simply vanishes over time as the molecules rearrange themselves. Optical storage has a longer life span but for critical archiving it might be necessary to get expert advice on storage variables such as temperature and humidity. CDs are very stable and the technology for writing to them is now affordable.

9.5 LATERAL THINKING

After a basic introduction to modelling I ask students how they would model a square, moulded picture frame with mitred corners. Quite correctly they create a 2-D template representing a cross-section through the moulding, extrude a length of moulding, clone it to create four lengths, rotate the lengths in units of 90°, and align the lengths. Some students use a Boolean operation to mitre the first length before cloning; some forget the mitring until after cloning; some ignore mitring entirely. They have created a frame the way they would if they were in the woodwork shop and the result is as required.

But how about creating the 2-D template (off centre) and then spinning it with the spin sides set to '4'? You arrive at the same thing with a lot less effort, but having used a process which does not mirror the real world frame-making method. After the superiority of this second method is acknowledged I ask how the students would change the frame from square to rectangular. A scaling operation along the appropriate axis is proposed but when we try it we find that whilst the frame is now rectangular, the moulding at the ends orthogonal to the direction of 'stretch' has itself become stretched (and is now fatter than the other pieces of moulding). Several alternatives then arise to solve this problem, but the exercise has been sufficient to show that thinking around the problem can pay off and that there are both advantages and disadvantages in making computers mimic real-world methods.

Consider also a simple cylindrical block. A cylinder can most obviously be an extruded circle, but what differences would the cylinder have if it were the result of a spun rectangle? The faceting of the end caps would be visible in wireframe view and might be more amenable to editing, but would this be an advantage or not?

Since the program would need (internally) to triangulate the end caps whether these were visible or not, would the file sizes be any different?

9.6 CHEATING

The shadows cast on the ground by the moving aircraft in flight simulators add a great deal to the realism of the scene, but look closely at the shadow and you will probably see that it doesn't follow the contours of the ground it pretends to be cast upon and might completely ignore the hangar it should be wrapped around. I have been told a number of times in computer animation houses that cheating is the name of the game – creating something as simply, quickly (and cheaply) as possible to fool viewers into thinking they have seen what you want them to have seen. Unless you have moral objections, and believe that the viewer is being cheated by shortcuts to achieve a believable result, then it is a good idea to join in and save yourself some work.

The philosophy of the movie world is not, however, appropriate to the engineer, who will usually need to construct accurately all elements of his model in such a way that they can be built in real life. He might, however, be able to 'economise' on the rendering without compromising the model's integrity. Consider the maxim that 'Efficiency is intelligent laziness'.

CHAPTER 10

THE FUTURE

The next generation of computer hardware and software is never too hard to predict. It will be faster, smaller and, after it has settled in the market, cheaper relative to its performance. It has often been said that it doubles in speed and halves in price every eighteen months. On arrival it will bring no great surprises since it will have been thoroughly hyped before it comes. Its performance will briefly seem incredible and it will raise the expectations and demands of users, thus preparing the ground for a further generation. In many areas of computing, standards are high enough for most tasks and one has to ask how much a superior technology is really needed. When writing this book, for instance, my workstation offers little practical improvement over the cheap, home micro I used for my first book; the software alone is worth nearly as much as my previous hardware and software combined and it is noisier, the bigger screen, smoother interface and better keyboard being the most obvious advantages.

When I come to view a model, however, there is no comparison; I repeated a ray tracing of a simple model (at the same resolution and colour depth) which had taken 27 hours on my first machine and it was completed in 72 seconds on my next. On my latest machine it would take only two or three seconds. This is the result of both improved and optimised hardware and of more efficient algorithms written to exploit the potential of the hardware. Now that I have more than 16,000,000 colours at my disposal instead of just 16, reproduction quality resolution, a high level of anti-aliasing, the ability to scatter light sources about a scene, add fog, image map, environment map, and make everything glossy, reflective, refractive and transparent, I am annoyed to find that my rendering times have gone up again. If anything tests a system's limits, rendering does, and it has become something of an Achilles' heel for the modeller. Although the requirement for high quality modelling and rendering is growing fast, and each generation of machines brings big improvements, it will be some time before you can move a light source in a complex scene with immediate, interactive, ray traced feedback. Meanwhile I quick-shade during the day and let the machine ray trace animations with all the bells and whistles during the night.

It is a less interesting exercise, therefore, to try and predict when things we can already do on our computers will be done better, and more challenging to guess what new feats we will be able to perform. Best of all is to wonder into what fresh modelling fields we might be taken.

10.1 VIRTUAL REALITY

The obvious uses for this infant technology are remarkable, the less obvious uses staggering, and despite the fact that it only reached the marketplace around 1989, it is expected to revolutionise many disciplines. We have already noted that surgeons are starting to practise convincing operations on 'virtual' patients, astronauts practise space manoeuvres on the ground, anyone can be a 'virtual' racing driver in his own living room, and animators can define an actor's path through 3-D space with a sweep of his hand. Whilst the ultimate video game scenario is one promise offered, there are many more potential uses for virtual reality, both mundane and outrageous.

Virtual reality has grown from concepts such as 'artificial reality', a term coined some years back by Myron Krueger, and is currently typified by its enabling encumbrances like the glove, mask and suit. The original DataGlove from Jaron Lanier's company VPL is an input device which is worn like a glove and translates hand and finger movements into electric signals. Combined with an absolute position and orientation sensor, the glove translates movements made by the operator's hand into information which can be used to duplicate the movements in the computer's three-dimensional world. It is thus possible to control movement within the computer scene by hand movement, and one obvious development has been to create a computer model of a hand which can mimic the operator's hand. It is then easy to create an object in the computer's imaginary scene and to grasp it with the model hand which inhabits that same scene, under the control of the DataGlove. The glove is also an output device, as tactile-feedback devices can give the operator the same touch clues as would be expected from manipulating a real object.

Whilst the glove can be used with a 2-D VDU screen to display what is happening in the scene, this falls short of providing the total control which participation in the 3-D scene would allow. The glove can therefore be used in conjunction with a stereoscopic headset which provides a separate screen for each eye and allows the user to look around the scene as he would in real life, presenting fresh views as it senses the head being moved and utilising all the spatial depth clues that the user would normally expect. This describes a fully 'immersive' VR experience where the user enters into the computer space, but 'non-immersive' VR is more easily provided where the operator watches a single screen to look around and to monitor movements he (or his hand) 'makes' in the scene. A further level of removal has the operator using external devices, such as the standard mouse, to control a proxy that has been modelled within the scene.

The gloves can, as you would expect, be used in pairs and a DataSuit has also been constructed to allow the whole body of the operator to interface with the machine. The most obvious limitation on the feedback from these devices at the moment is the lack of force-feedback. It is possible to feel the surface of an object, but not to feel its weight when 'picked up'. More effective force-feedback devices have been built but not yet in a convenient form to match the relative freedom of the Data Glove/Suit/Headset. Forces and torques can be applied to a hand control but only as part of a substantial machine rather than in the compact and mobile form required, though the possibility of using 'memory metals' to push against the skin in the DataGlove has been considered as a response to the force-feedback limitation.

The technology of this new discipline seems to define little more than a fresh interface paradigm, but VR has somehow been presented as a branch of philosophy. However, now that the pioneering enthusiasm has become either more muted or more local, it has become popular to tone down expectations by use of the term 'virtual environment' (which accurately describes the computer-generated space), rather than 'virtual reality' (which seeks to describe the experience of using it). Either way, the 3-D environment built in the computer is a prime example of modelling and the pressing demands made by VR (even with its current, relatively crude, technology) must impact strongly on the development of modelling and rendering.

The ability to move through a computer scene uses nothing that is new in principle, it merely requires a sequence of views of the scene from different viewpoints. The ability to provide this is a necessary part of any 3-D modeller but in order to give a sense of movement these views need to be presented to the observer at about 25 or 30 views per second (these examples being the standard video animation rates in the UK and USA respectively). If your machine can't render a complex scene in 0.04 seconds then you might need to export your files to one that can.

Whatever the object you are modelling, however, whether it is an office block, a car, a container for washing-up liquid or a teapot, it is likely that you can enhance your (and your client's) understanding of it by experiencing it in three (and four) dimensions. This is one thing that VR offers you. You will also be able to do much more comprehensive pre-production testing in VR than is currently possible, perhaps using yourself as the model for ergonomic analyses: walking through the office, sitting in the car, handling the bottle, pouring from the teapot.

10.2 SHARED ENVIRONMENTS

Recent developments on the Internet use VRML (Virtual Reality Modelling Language) to define 3-D scenes in such a way that they can be accessed by any user. Early versions defined only the scene but later versions allow for description of behaviour within the scene. This is subject, of course, to the user's browser (Internet access software) being configured to handle the language. A user can then 'explore' the site's virtual environment or a number of users can then 'share' it, even being able to interact with one another if the system is so designed. Early models are relatively crude, limited by the current bandwidth of the Internet, but provide the basis for visions of communal design input to a virtual model.

When a user enters such a virtual space either the scene can be 'seen' normally or a proxy of the user, known as an avatar (from Hindu mythology), can be added to the scene and viewed like an 'out of body' experience.

10.3 DYNAMICS

Increasingly it will become possible to give to models properties that go beyond the merely visual. At the moment in a model, the materials comprising a rubber tyre, a steel girder and a silk handkerchief are likely to be differentiated only by their surface appearance. By crediting materials with the attributes that they have in the real world, it is possible to have them reacting to physical forces and interacting with one another and with their environments.

Alan Barr (*Introduction to physically-based modelling*, SIGGRAPH, 1989) defines a physically based model as 'a mathematical representation of an object (or its behaviour) which incorporates forces, torques, energies, and other attributes of Newtonian physics. With this approach, it is possible to simulate realistic behaviour of flexible and rigid objects, and cause objects to do what we wish them to do (without specifying unnecessary details)'. In section 6.5 it was explained that, having defined a ball and a surface, we can 'drop' the ball and watch it bounce around until coming to rest. If we also care to drop a properly defined cup onto either a properly defined pillow or onto properly defined concrete, we can witness it either land gently or smash at a level of accuracy prescribed by our definitions. It is immediately apparent that many objects are flexible to different degrees, either in whole or part, and that many are articulated (such as a skeleton). When this is mathematically acknowledged in a model, however, snooker balls can interact realistically, articulated figures can trampoline and leaves can gently flutter down in a breeze. Not surprisingly, some things are easier to define than others and many things are too complex for it to be practical to attempt a definition. Nevertheless, in addition to the functionality of such a display, there is always an addictive magic about witnessing even a simple occurrence unfolding under its own (apparent) initiative.

The application of basic physical laws enables the realistic simulation of the motion of bodies and cannot only be applied to separate complex bodies, but can automatically describe the conduct of bodies in collision. It therefore takes little effort to simulate a raindrop falling or a ball being thrown by specifying mass, starting velocity and direction to a system that knows what gravity is, perhaps that there is a cross-wind and how to apply the rules. A little more information is needed to cope with friction and bouncing and more again if the object is articulated, flexible or asymmetrical; but although the number and length of the required mathematical equations grows, the governing rules remain clear. There is a considerable body of literature on the subject and it has been a major conference topic for several years.

10.4 INTELLIGENT MODELS

A term recently imported to computer graphics (from the fields of philosophy and biology) is that of 'teleological modelling'. Derived from the Greek word 'teleos', meaning end or goal, it provides an extension of the current definitions of modelling to include a number of recent developments and provides a model that is goal-orientated. It is a mathematical representation which calculates the object's behaviour from what the object is 'supposed' to do. Alan Barr suggests that it has the potential to extend the scientific foundation of computer graphics and vastly to extend the state-of-the-art for computer graphics modelling. He suggests that teleological methods can create mechanistic mathematical models with predictive capability and produce compact formal descriptions of complex physical states and systems. It seems that teleological modelling does not offer new methods but provides a conceptual framework within which recent (and future) methods governing an object's purpose can be related to existing modelling methods. The teleological model of an object includes time-dependent behavioural goals as part of the object's fundamental representation. This gives us an object that knows how to act (perhaps a doctor's 'model' patient can bleed?).

As well as knowing how to respond physically to the world, and having an understanding of its role in life, it might be useful for an object to be able to exhibit intelligence. It could then recognise imminent collisions and take avoiding action, plan routes and strategies, and respond to rules of behaviour. Artificial intelligence (AI) has been described as the science of making machines (or, in our case, models) do things that would require intelligence if done by humans. It is a new and important science, whose methods need not concern us here but whose influence will be felt in modelling. Already 'expert' or 'knowledge-based' systems can be used within limited domains, such as architecture, to allow models to be interrogated for hidden information; the system's knowledge having been initially collected from human experts.

It is also possible for a system to develop heuristically, i.e. to learn from its own experience. Paraphrasing Norbert Weiner (from his seminal book *Cybernetics*) an object that learns 'is one which is capable of being transformed by its past environment into a different being and is therefore adjustable to its environment within its individual lifetime'. We might build models that develop the initial properties we give them through their own experience and 'children' of these objects could, of course, inherit their more advanced properties. Each generation would build on the understanding of its ancestors.

Systems already exist which exhibit a behavioural response to their world, an elegant example being Craig Reynold's flocking model birds. He describes a flock as being made up of discrete birds yet with an overall motion that seems fluid; it is simple in concept yet is so visually complex; it seems randomly arrayed and yet is magnificently synchronised; and perhaps most puzzling is the strong impression of unintentional, centralised control. In his first models the birds (he calls them 'boids') are simplified to wedges, and by applying a few simple rules a group can 'fly' from a starting point to a given goal, avoiding obstacles en route and displaying all the characteristics of a flock. The wedges were later replaced in an animation by bird-like models which could flap their wings in flight and also display the flocking characteristics.

The behaviour of an object can be described in terms of its response to stimuli within its environment and that response may be qualified by its internal state (if it is credited with one). Several research projects have set up computer models inhabited by 'creatures' that mimic internal states such as hunger and anger and use these states to modify goal-directed behaviour. We can, therefore, think of future models having properties that stretch from red and glossy to angry and confused.

10.5 DIGITAL DOUGH

For some time I have been awaiting the arrival of what I think of as 'digital dough' which would permit the creation and manipulation of computer models as a sculptor does with clay. This would assemble several existing principles to create a new, interactive modelling medium. If a three-dimensional lattice was created to define an object then the object could be deformed by moving points in that lattice (the 'resolution' of the object being proportional to the closeness of the points). If the points were interconnected with springs and dampers then the deformations would transmit through the solid, and the consistency of the object could be defined by the tension of the springs. If we now let this object exist in a virtual world accessible through technologies such as the DataGlove, then it can be hand-modelled like clay; a sort of digital dough. By putting on the gloves we can squeeze, stretch and shape the object like a sculptor and, by changing the tension of the springs at any time, can change the object's consistency. This becomes a much more intuitive modelling method than those with a more visible mathematical basis.

Taken further, an articulated human figure model could be constructed with parts made of digital dough, thus providing a lay figure which can be hand-modelled to suit any requirements. We can make it short and fat, tall and skinny, with a big head or large feet, can create caricatures or likenesses, and could easily 'tweak' the quantities of dough available in any particular area. Once sculpted to taste, the figure can be animated using all the normal techniques, including dynamics, but the glove technology also offers the option of combining them with interactive positioning. The figure could be moved like a puppet, set in position by hand and/or controlled by a program, and given whatever degree of intelligence we choose. We can define characteristics such as gait pattern with 'conventional' methods or by real-time demonstration using the figures themselves (the figures have become plural, because it is, of course, trivial to clone a crowd). Facial expression can be similarly controlled and artificial intelligence can be attributed as desired. Its proportions could also be entered numerically to ensure the accuracy needed for scientific testing. If we don a Data Suit (or launch an avatar) we can join our new creation in his/her digital world.

10.6 CATALOGUES

Alvey Ray Smith suggested in *Byte* magazine (Sept., 1990) that 'much modelling will be made redundant by the selection, from electronic catalogues, of ready-made models. The user becomes a "spatial editor" inserting the model(s) into the 3-D scene, sizing and customising to taste'. In a very small way this is already happening with 3-D clip art, but it takes little imagination to see it developing. Builders already design houses by amending plans from house catalogues and 3-D model catalogues potentially avoid similar objects being endlessly remade, though at some threat to originality. It is increasingly common for clients to be able to have customised kitchen designs created for them as they watch by designers selecting, modifying and arranging a collection of kitchen components.

Conversely, photo realistic models of items that do not exist can be shown in catalogues, and not produced in reality until ordered. One jeweller has for some years used a 3-D modeller to design and sell jewellery before it has been made with obvious savings in outlay on precious materials, also gaining the ability to tailor items to individual demand.

The Internet offers a massive global catalogue and sites are already springing up to offer free and 'shareware' models in standard file formats. Commercial sites have a growing range of models available at varying levels of detail, all with the option of immediate downloading and electronic billing. These may be immediately suitable off-the-shelf for use in many contexts but also offer the basis for editing to suit more specific ends. Indeed it may become more common to edit bought-in models than to create originals from scratch. The wheel need not be reinvented, even if there is a price tag attached to the convenience. A further consideration is that, since the Internet allows simultaneous access to both client and designer, new client/designer relationships will evolve.

10.7 AND FINALLY

The magnitudes of forthcoming changes are unpredictable but looking at the pattern of the past forty years suggests an exponential growth curve in computer modelling as in so many other areas. Pocket-sized workstations might not get given away free with petrol but they will become commonplace. This resultant increase in portability will not just add to convenience but will change attitudes to computers – an ever-available, hand-held device losing the preciousness of a desk-bound machine available only during office hours. Computers will also become increasingly an embedded technology and, as such, invisible (but everpresent?).

3-D digitisers will become automatic and commonplace. Solid modelling will be used more (it is already appearing in computer games). Visualisation, such as medical and financial, will increasingly use 3-D models to make otherwise obscure data visible. New modelling methods will evolve to deal with problematic subjects such as freeform, non-geometrical, soft objects (e.g. humans); objects for which there are no straightforward mathematical descriptions.

Increased access and bandwidth on the Internet will move us towards the realision of long-held visions of electronic, global communities and will encourage home working from 'electronic cottages'. Virtual reality will blur the distinction between the world of the model and that of ourselves. Modelling is exciting already but it will blossom in the next few years.

CHAPTER 11

3D STUDIO (FOR DOS)

11.1 INTRODUCTION

Autodesk 3D Studio is a sophisticated computer graphics and animation package for the PC. There are a number of versions and releases of 3D Studio. This chapter will deal with 3D Studio Release 4 for DOS. This is a very widely used, well established, general modelling and animation package. Other products from Kinetix (a division of Autodesk) are 3D Studio MAX (which came out after 3D Studio, is now on its second release and is widely used by animators and game developers) and 3D Studio VIZ (the most recent product, aimed at architectural, civil and industrial designers). MAX and VIZ both have a greater hardware requirement than 3D Studio and the interface is completely different. MAX and VIZ will be looked at later in the book.

If 3D Studio is used in the text, it will be referring to 3D Studio releases 2, 3 and/or 4 for DOS. 3D Studio MAX and 3D Studio VIZ will be called MAX and VIZ.

11.1.1 HARDWARE REQUIREMENTS

The realistic minimum hardware requirement for 3D Studio (Dos) is a 486DX processor or newer, 16MB of RAM (increasing to 32mb is an advantage) and about 100mb of hard disc space. A good 17"monitor is useful for more complex work. 3D Studio runs best directly through DOS. It can run through Windows but this is not recommended.

As with many expensive software packages, 3D Studio, MAX and VIZ need a hardware key or 'dongle' to run. This is shipped with copies of the programs and is a small box which fits into the parallel port at the back of the PC. If this dongle is not present, or the system for some reason can not find it when it boots up, the program will not run.

11.1.2 AN OVERVIEW AND SOME BASIC PRINCIPLES

3D Studio has five modules which make up the program. The **2D Shaper**, the **3D Lofter**, the **3D Editor**, the **Keyframer** and the **Materials Editor.** They are all linked and are like independent, simultaneously running, programs. You move from one module to the other, developing aspects of form creation, form manipulation, materials qualities, lighting, scene building and finally, rendering or animation. The 2D Shaper is used to create, or import, two dimensional outlines which can then be 'swept' along a path in the 3D Lofter to create a huge variety of forms. These objects can be further manipulated, assembled and added to, to make more complex objects and scenes in the 3D Editor. Also in the 3D Editor textures and colours are added to the objects, the scene 'lit' and camera views defined. The Materials Editor is used to create and manipulate surfaces and colours and finally, if desired, the Keyframer can be used to create an animation.

3D Studio MAX and VIZ, while sharing many conventions, have different interfaces and do not have the five modules.

The modules are accessed through the Program pull-down menu or, more quickly, through 'hot keys' F1 to F5. The 2D Shaper, the 3D Lofter, the 3D Editor and the Keyframer, because they 'pass on' and 'share' geometry, all have similar interfaces, with a number of common pull-down menus, side menus and an Icon panel. The 2D Shaper has one window or 'Viewport'. In the Lofter, 3D Editor and Keyframer, the object or scene can be viewed in a series of Viewports. The nature of these Viewports can be defined, for instance most commonly used are top, front, side and camera. This gives a good understanding of the objects and their orientation to each other in the scene.

The pull-down menus are similar across the modules (with the exception of the Materials Editor). They are grouped under the headings of Info, File, View, Program and Network. The Info column gives you information about the program, its status and allows you to access some of the most used setup parameters. File contains all the saving and loading commands. Views is not only where the viewports are defined, but contains various other commands that influence the display. Program gives access to the different Modules in 3D Studio, and Network is only used if you are using a number of networked machines for rendering. Each module's pull-down menus do have some commands which are unique, and while some of the functions will be explained in detail in later chapters, they are best explored and used to gain some familiarity with them.

The side menus are used in a similar way across the modules and almost always follow the convention of giving a command in stages. For example to create a sphere in the 3D Editor, you pick create, then G-sphere, then smoothed. Each step of the command remains highlighted and the final instruction appears at the Command Line at the bottom of the screen. If you return to the create command, the last command given will still remain highlighted. This saves time by not having to 'click' down through the stages of the instruction again.

In the lower right hand corner of the 2D Shaper, 3D Lofter, 3D Editor and Keyframer, is the Icon Panel which contains several push-button icons. These are used mainly to adjust the view in the viewports, and once again many are common to all four modules (not the Materials Editor). The Universal Icon Push-buttons are Pan, Zoom In, Zoom Out, Zoom Extents, Zoom Region and Full-Screen Toggle. The Selected, the Hold and the Fetch buttons are also found here and their significance and use is dealt with later.

11.1.3 GRID AND SNAP

Grid and Snap keep the drawing accurate and are available in all the modules (except the Materials Editor) and should be used most of the time. These features are found in almost all CAD packages. The Grid is a network of regular dots which can be used for reference while moving segments, vertices, polygons or shapes around. The Grid can be displayed, or turned off, and set to any appropriate scale.

When Snap is on, the mouse action will 'snap' to evenly spaced fixed points, ensuring accuracy. It can be set to any appropriate scale for the job in hand and can be set independently in each axis. When using and displaying a Grid, it is best if the Snap relates to the Grid; for example if the Grid is set to 10 the Snap should be set to 10, 5 or 2. Snap is an important feature if working to accurate dimensions. Often users change the Snap settings several times during a modelling session. If, for example, an object is to be moved a fixed known distance, it is sometimes best done by setting the Snap to that distance, moving the object, and then resetting the Snap to its original working setting.

There is also an Angle Snap. This varies the angular step through which objects or polygons can be rotated about an axis and can be set from 0-360°.

11.1.4 UNDOS AND DON'TS

In 2D Shaper and 3D lofter there is an 'undo' button. This will cancel the last action. Unlike some programs, there is no multiple undo in 3D Studio. Those of us who are used to using programs like Autocad, and have become accustomed to counting our way back through a series of commands to the last mistake, need to beware! With 3D Studio, the commands Hold and Fetch need to be used regularly. Hold temporarily saves the work in the module at that point, and Fetch returns to that point. Their use is particularly important in the 3D Editor and the Keyframer as, due to the amount of data involved, there is not even a 'one step' undo!

11.1.5 THE AXIS

All the modules share a common Axis. The default or Global Axis is centred on 0,0,0 of the 2D or 3D space. The position of the Axis can be modified. It can be placed at any point in the view, can be aligned to a vertex, a face, an object or group of objects. Its position is particularly important in modify commands such as Scale, Rotate, Skew and Taper.

There is also a Local Axis option. This command centres an axis on any selected geometry. This could be, for example, a single polygon, number of polygons, group of 3D objects or a selected group of vertices. If a single polygon or shape is selected, the Local Axis will be placed exactly at its centre and any alterations will be based around this point. If the Local Axis is inactive it will rotate about a Global Axis outside itself. The results are particularly striking if the Axis is changed when a Selection Set of vertices is made. The figure on the right demonstrates the effect of using a Global Axis and Local Axis on some 2D Shapes and selection sets of vertices.

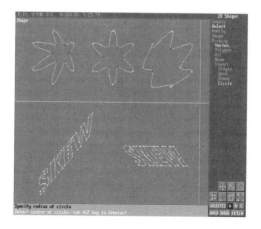

*Affect of a local and global axis
on some 2D commands*

11.1.6 THE GEOMETRY

As with many modelling packages, the 3D objects that 3D Studio creates, are really only hollow shells and do not exist as 'solid' geometry. The sole building block of any 3D object is the Triangular Face. All the surfaces in 3D Studio are built up out of a series of these triangles. For example a cube will contain twelve triangles, two on each face, while a smooth looking sphere may contain hundreds. The ends of the triangles, or vertices, are what the computer uses to orientate the position of the various parts of the object in an infinite 3D Space, using X, Y and Z coordinates. 'Hung' on these vertices are faces which in turn have three edges. It is by building up groups of these triangles, that 3D Studio creates objects.

3D Studio's ability to smooth the intersection between each triangle so that seams do not appear at rendering, means smooth curves are possible. The density of the network of triangles on smooth shapes can be important if the illusion of smoothness is to be achieved. The figure to the left illustrates this. The object on the left is a low density sphere with one hundred triangles or faces while the high density sphere on the right, has 1,600. At smooth rendering (phong or metal) the right sphere is smooth and ball-like, whereas the low density sphere, while smoothed in the middle, displays the facets of the triangles on the edge.

Objects may not appear to be constructed in this way, as not all the shape details (i.e. all the triangles and edges) are shown in the mesh images displayed on the scene. This is done as a default by the program, to make the display less cluttered and to speed up the time it takes for each

Rendered low density and high density spheres

viewport to be refreshed. The complexity of the default display can be altered to show more or less detail. When all the vertices, edges and faces are displayed in a complex scene, it is apparent just how much information the computer is having to calculate. This highlights one of the most important issues in the building of models for rendering and animation; the number of vertices and faces should be kept to the minimum needed to achieve the desired rendered result. A model should always be built in the least complex way possible. The more complex the model, the slower the rendering and the larger the files. There are many ways in which the complexity of the models can be reduced and a good modeller finds ways of reducing the complexity, without sacrificing quality in the final rendered image. The figure on the right gives a basic example of this. The 3Ds are of different complexities and vary considerably in the amount of numerical information which is contained in each. However, as we can see when they are rendered, they all look the same.

11.1.7 SAVING FILES

3D Studio saves the information in each of the modules with several file formats. SHP files are used by the 2D Shaper, they save all the 2D information in this module. LFT files store the contents of the 3D Lofter. 3DS files contain all the data of the 3D Editor and the Keyframer-the meshes and their material assignments, all lights, cameras and animation information. This is a useful format to use as any meshes that are stored can be merged into other scenes in the 3D Editor. MLI files contain the material definitions, that is the settings and the names of image maps used, and are the format used in the Materials Editor. The best format to save work in is a PRJ or Project File. This saves all the information in all the modules. This is the format used for 'work in progress' as it makes it possible to return to the point where work was terminated in every module.

The 3Ds vary considerably in complexity, but when rendered, they all look the same

11.2 THE 2D SHAPER

The 2D Shaper is for vector based drawing. It is used to create and manipulate 2D Shapes and is similar to, and compatible with, a number of other 2D programs. 2D shapes can be imported from programs such as AutoCAD, Coral Draw, Adobe Illustrator and Mini CAD, using common file formats such as DXF. Text can be created in its own internal Text Editor or it will accept any Post Script font.

11.2.1 THE INTERFACE

The means of giving commands and accessing functions in the 2D Shaper follow the conventions that have already been discussed in the Overview. Along the top are the pull-down menus that give access to the Files (opening, saving etc.) the Views, the Snaps and the Programs. The side menus are used to give most of the drawing and manipulating commands; select, modify, shape and display all have their own sub-menu of commands. These use the convention of the 'layers' of instruction. For example to erase or delete a polygon the command 'line' is Modify/Polygon/Delete. This convention is used throughout the Program. There is also the Icon Panel with a variety of zoom buttons, a Selected button and the Hold and Fetch Buttons.

11.2.2 POLYGONS

All lines and shapes in the 2D Shaper are referred to as Polygons. A Polygon in the 2D Shaper is made up of two or more Vertices (or control points) with a Segment between each. The 2D Shaper can create almost any 2D shape. Any shape that cannot be made in this part of the programme, can be imported from another program. The most basic unit in the Shaper is the line. By clicking with the mouse (or by clicking, holding and adjusting while the line is being drawn) straight or curved lines can be drawn, or a combination of both. Other commands are Arc, Quad (squares and rectangles), Circle,

Ellipse and n-gon (multi-sided regular polygons which can have flat or curved sides)

3D Studio uses the terms 'closed' and 'open' to define the polygons. The closed polygon (one in which the end points connect so the shape is a closed loop) is most commonly used to define the cross section of an object at any point as it is lofted (often called sweeping or extruding) along a path. The open polygon (or line) is most often used as the path along which a closed polygon (or form) is lofted.

While it is possible to create almost any 2D form in the Shaper, all shapes intended for lofting must be closed polygons and must not self inter- sect. More than one polygon can be lofted at a time as long as it does not overlap another. For example, two non-overlapping circles can be lofted to make a 3D tube. This can be very useful if, for example, you wish to make a flat shape with several differ- ent shaped holes. This is preferable to using a '3D Boolean' operation, explained later, to create the same effect as it is quicker and the final model contains less numerical information. The top right image shows usable and unusable shapes for lofting. The other three images illustrate multiple shapes being lofted.

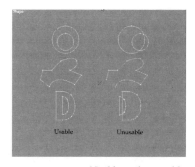

Usable and unusable shapes for lofting.

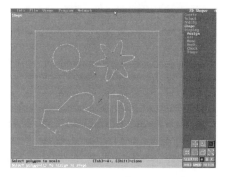

Creating outlines in the 2D Shaper

The finished object

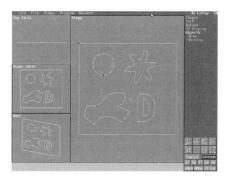

Lofting the outlines in the 3D Lofter

11.2.3 ADJUSTING POLYGONS

The basic 2D building blocks, open and closed polygons, can begin to be manipulated creatively to create complex forms. The Move, Rotate, Scale, Skew, Mirror, Adjust, Linear, Curve and Delete commands give a variety of manipulation options. The directional cursors (selected with the Tab key) can be used to restrict movement to horizontal or vertical. This can be very useful for keeping objects aligned. Several of the modify commands such as Rotate, Scale and Skew, are influenced by the position of the Axis in the scene.

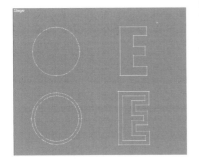

Polygons can be 'cut up' by deleting vertices or segments, and then joined together again using the Weld command. Creating closed polygons, cutting them up and welding them back to together, is a very good way of creating closed shapes in 3D Studio and is used in several of the modelling examples later in the book.

The outline tool works on already created shapes and draws an outline of the polygon. If a line is used, it draws a closed shape that surrounds the line at a fixed distance. On a closed polygons, a smaller and larger outline will be created at the defined distance.

The 2D Boolean tool acts on overlapping shapes and combines them in one of three ways: union, subtraction or intersection, which is chosen through a dialogue box. This dialog box also offers the opportunity to weld the polygons which, after the operation, ensures that you have a closed form. Using Hold is a good idea as the original geometry will be destroyed in the Boolean operation.

2D Outline

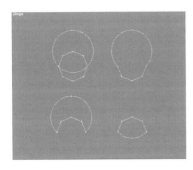

2D Boolean

11.2.4 SPLINING

The Vertices in the Shaper are not the same as those used to describe forms in 3D. They are only in a 2D plane and each Vertex (or control point) is capable of being adjusted to define the curvature of the lines that enter it. These types of lines are called Splines and enable complex curvy lines to be defined by relatively few vertices. Each individual vertex (or selected group of vertices) can be moved or have its spline adjusted. By picking a vertex and holding and moving the mouse, two directional arrows appear. The red arrow represents the incoming tangent vector and the yellow arrow the outgoing. These vectors and be altered together or independently.

Examples of 2D splining

11.2.5 FLAT SHAPES

Sometimes it is desirable to have 2D flat shapes in a 3D scene. These begin life in the 2D Shaper. In the 3D scene they will have no obvious thickness and will disappear when viewed from the side. This may be useful if making, for example, a sheet of paper. However it is more often used as decals and lettering placed in front of larger 3D object. A sign on a wall would be an obvious example. The same rules apply to flat shapes as to shapes for lofting, i.e. they must not self intersect or overlap and they must be closed. Where possible flat shapes are used in place of lofted shapes. Lofted shapes contain more vertices and hence increase the complexity of the drawing. One of the modeller's objectives at all times, should be to keep objects and scenes as simple and with as little numerical information as possible in them, to create the desired effects.

Text creation is achieved through the 2D Editor's Text Editor. There are a number of pre-loaded fonts which come with the programme and hundreds of additional fonts are commercially available. The text to be input is typed in and placed in the shaper with a 'clink and drag' box. The text can be input in any width and height ratio, however the aspect ratio of the original font can be retained by holding down the Ctrl key while drawing the placement box.

11.2.6 PREPARING A SHAPE FOR LOFTING

Once usable shapes have been created, imported, combined and/or manipulated, they will often be used for lofting (or sweeping) in the 3D Lofter. This may be a straightforward 'thickening' by extruding along a straight path, or something much more complex.

So that the next module, the Lofter, knows which polygon(s) to import, one or more have to be assigned. Assigned shapes turn yellow and the Shape/Check command verifies their suitability for lofting. If suitable they can be imported, if not they have to be reworked. The most common problem that makes a shape unusable is that one or more of the vertices, which appear on top of each other, are in fact not joined. By using a combination of selection sets of vertices and a Vertex/Weld command, this problem can usually be solved. This is a very common problem with shapes imported from another program.

11.3 THE 3D LOFTER

11.3.1 LOFTING

The basis for much of the modelling in 3D Studio, as it is with a number of other 3D modelling packages, is the 'sweeping' or 'lofting' of a 'shape' along a 'path'. The 3D Lofter uses the polygons created in the 2D Shaper to do this. For example, a circle lofted along a straight path produces a cylinder. However both the path, and the shape to be lofted, are infinitely variable. They also can be altered, changed or deformed as they move along the path, varying the cross section and angle at any point. All this gives enormous modelling potential.

There are a number of rules which have to be adhered to with both shapes and paths. For example, Shapes (discussed in the earlier section on the 2D Shaper) must be closed polygons and must not self-intersect. The position of the shape relative to the path can have a significant affect on the resulting form, particularly if you are using a Deformation Tool. The usual procedure is to centre the shape on

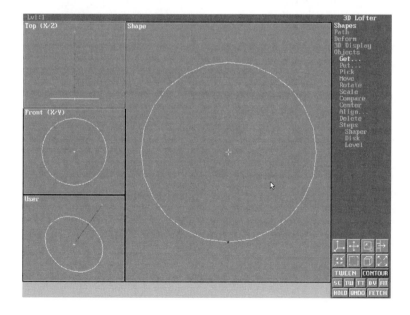

The 3D Lofter Interface

the path and this usually gives a predictable result. However altering the position of the shape relative to the path, can open up new modelling opportunities. This is exploited in some of the case studies later in this section.

Paths can be more complex than the shapes to be swept or lofted along them. They can be open or closed polygons with any number of vertices, they can also self-intersect. They are usually created and imported from the 2D Shaper, however other suitable paths, both flat and three dimensional, can be imported from compatible programmes, a DXF file from Autocad for example.

Along the path, each vertex in the line that was used to create it, is marked with a dash. The default path in the 3D Lofter is an open straight line, one hundred units in length, with a vertex at each end. Between each vertex there are a number of ticks, called steps. When lofting a shape up a path, a copy or version of the extruding shape is 'put' on each vertex. By introducing extra steps, the smoothness of the loft can be refined, interim cross sections being placed at each step. The number of steps can be set to between zero and ten. However, increasing the number of steps also increases the models complexity, so only the number of steps needed to achieve the desired results should be used. A straight extruded shape for

A circle lofted up a helix path

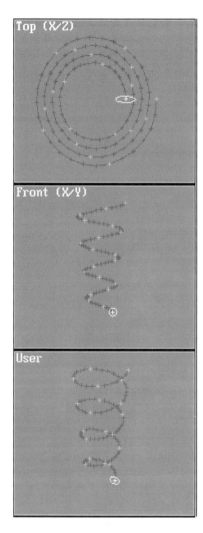

A helix path created by the 2D Lofter

example, needs no interim cross sections as its shape remains constant the length of the object. In this instance there should be only a vertex at each end and the steps setting should be zero. A more complex curving shape may need lots of vertices and lots of steps to achieve the necessary detail.

As well as importing paths, there are two types of path which can be created automatically in the Lofter, Helix and SurfRev. The Helix option makes a path in the shape of a helix, shown, where you can define the start and end diameter, the height and the number of turns. The Surface of Revolution (SurfRev) command creates a circular path or a proportion of a circle. This is often called a 'lathe' command. Objects such as bowls, plates and vases, where you create a profile and 'spin' it around an axis, are often modelled in this way.

11.3.2 LOFTING DIFFERENT SHAPES ALONG A PATH

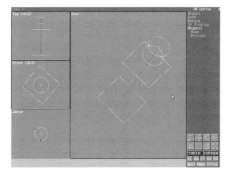

All the vertices and steps represent 'levels' in the path and each can have a different shape assigned to it. The default 'working' level is at the bottom of the path and any imported shape will be inserted there. However Page-up and Page-down move the active level on the path, allowing shapes to be imported and/or altered at each level. Because the Lofter simply connects one vertex in one shape to the same vertex on the next shape at the next level, each shape on each level must have the same number of vertices. This is usually achieved in the 2D Shaper by basing all the shapes intended for use at the different levels, on the same basic polygon which is copied and altered to create the variety of desired shapes. The bottle example shows different shapes being used along a path to create an object.

Different shapes being used along a path to create an object

11.3.3 DEFORMATION TOOLS

The 3D Studio Lofter uses a series of powerful Deformation tools which can be used to vary the shape as it moves along the path. They are Scale, Twist, Teeter, Bevel and Fit. Each use a similar methodology where a Deformation Grid is used to define how that tool will affect the final 3D form. Each grid is like a graph and share several conventions. White lines are the graph lines for the extent of the action, the yellow lines represent the levels of the path in the Lofter and the blue line is the deformation line. Most of the adjustments of this blue deformation line are the same as for editing a line in the 2D Shaper, where vertices can be created, moved and splined.

Of course the scales each grid use vary. For example, the Scale Deformation has percentage across the top representing the different scale at each point where the Deformation Line crosses a Level, while the Twist Deformation will have degrees showing the angle of rotation at each point along the loft. All the Deformation tools also allow you to define the deform symmetrically, or along the X or Y axis independently. More than one Deformation tool can be used on the same loft. Because they follow similar conventions, these very flexible and powerful tools are easier to master than they first appear. MAX and VIZ have the same tools.

Bevel Deformation

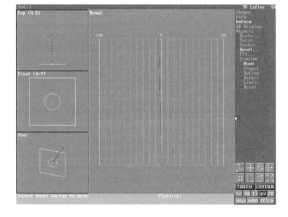

There is one Deformation tool which does not follow the conventions of the others, and that is Fit. It works on the principle that you can assign a profile in both the X and Y axis. When the shape is lofted, it is 'forced' to fit the profile in one or both axes.

The following illustrations demonstrate the basic Deformation tools. More complex examples of how to exploit a number of these tools, are demonstrated in the later case studies.

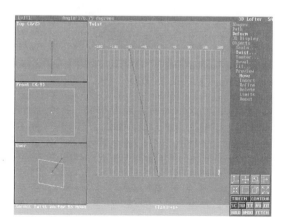

Scale Deformation

Twist Deformation

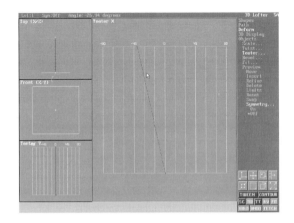

*Scale and Teeter
Deformation used
together*

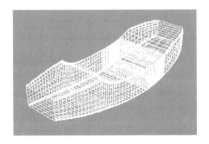

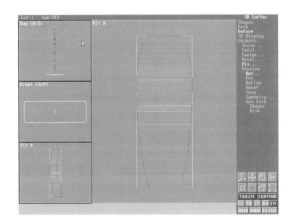

*With Fit Deformation, a shape is
assigned in the X and Y axis to
define the shape*

11.3.4 PREVIEWING AND CREATING

Once Paths have been created, Shapes imported and any Deformation Grid defined, an object can be lofted. Before making an object in the Lofter, it is always useful to use the preview facility. This gives an idea of how the object is going to loft by displaying a simple version in the display. It is particularly useful to help to predict the affect a Deformation Grid is going to have on a loft.

The command Objects/Make brings up a dialog box, see the figure on the right. The default settings are usually the most appropriate, however there are options offered to adjust several commonly changed parameters such as Cap Start/End (used to create a surface at the start and end of the loft), Smoothing (determines whether the object is smoothed for rendering) and Mapping (where mapping coordinates can be applied to the object as it is being lofted).

When Tween is on, the Lofter puts a cross section at each Step in the Path and not just the vertices as it does when it is off. Tween is usually needed when using a Deformation tool. When Contour is on, shapes are lofted perpendicular to the path. When inactive, shapes are lofted parallel to the X/Y plane. The figures below demonstrate a loft with Contour on and Contour off.

Object Loft dialog box

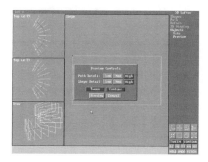

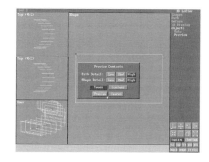

Lofting with Contour on and Contour off

Once the object is given a unique name, it is created and put into the 3D Editor.

11.4 THE 3D EDITOR

The 3D Editor is an empty 3D space into which the objects that have been created in other modules (or imported from other sources) are placed. These objects can then be manipulated and assembled to build scenes. It is in the 3D Editor that objects are assigned their material qualities and lighting and cameras added. We can imagine the 3D Editor as a stage or photographer's studio ready for some creative input. Most time is spent in the 3D Editor when using 3D Studio, particularly if still images, rather than animations, are being created.

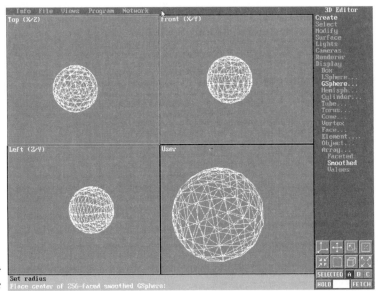

The 3D Editor interface

11.4.1 THE INTERFACE

As the scene is 3D, and the computer screen is 2D, views are set up in the 3D Editor which interprets the scene, and the orientation of the objects within it. The default setup is similar to that of the Lofter. It offers a Top(X/Y), Left(Z/Y), Front(X/Y) and a User view (an isometric or flat view). This setup can be changed by choosing Viewports in the Views pull-down menu, but for most modelling the default setting is fine. Also, once they have been created, a Camera view or views can be introduced. Unlike the User view, the Camera views are in perspective and give a real world impression of the scene. The 3D Editor's interface is most like that of MAX and VIZ.

11.4.2 THE GRAPHIC COMPONENTS OF THE 3D EDITOR

Within the Editor, objects are displayed as meshes of which the main building block is the Triangular Face. Each triangle is made up of a Face, three Vertices and three Edges. An Element is one of two or more mesh objects grouped together into a larger object. For example, if two side-by-side objects are lofted in the Lofter, a single named object is created with two Elements. Also when combining multiple objects into one object in the Editor, each of those objects become an element of a larger object. The object may be a car and the wheels be elements of it.

11.4.3 DISPLAY

In the 3D Editor, Objects, Elements or Faces can be displayed or hidden. Several Objects, for example, can be hidden while work takes place on others. Meshes can be displayed at several levels of complexity. Full Detail display shows all the graphic components, Vertices, Edges and Faces, and is the most comprehensive display (but of course the slowest to refresh). Box Detail, where the objects are displayed as a 'bounding box', is by far the quickest but is little use when making detailed alterations to meshes or scenes. Between these two extremes there are a variety of options. In some only edges are displayed, in others only the faces that point toward the viewer are seen. Unlike MAX and VIZ, 3D Studio can not have interactively rendered viewports.

When work is complete on a mesh and no further modifications or movement within the scene are intended, the object can be Frozen.

155

This causes it to turn grey in the display and it is ignored by the 3D Editor. This is helpful as the object can still be seen and worked around, but not accidentally modified or deleted.

11.4.4 OBJECT COLOUR

All objects in the 3D Editor can be displayed in any of 64 colours. The default colour is white, but when a new mesh is being created, it can be assigned a colour. A colour can also be assigned or changed in the 3D Editor. Using colour to group objects rationalises a scene, and makes the selection of sets of objects much easier. For example, all the furniture in an interior scene could be assigned a colour and different ones used for the structure and light fittings. To render the interior without any furniture, it could be selected using Select/Object/Colour and hidden to temporarily remove it from the scene. Any colour used on a mesh object in the 3D Editor has no affect on the final rendered object.

11.4.5 SELECTION SETS

A Selection Set is a temporary collection of one or more graphic components. Within the 3D Editor the use of Selection Sets play an increasingly important role, and learning how to master their use is vital. In the 2D Shaper vertices and polygons can be selected. By giving a command (and clicking the select button in the Icon Panel) you could affect a number of polygons or vertices with the same action. In the 3D Editor, Objects, Faces, Vertices and Elements can be selected. In a crowded scene, this can get visually complex. 3D Studio offers a number of ways of rationalising both the selection sets and the display with the use of colour.

11.4.6 3D EDITOR MODELLING

Explore any 3D program and you quickly find that there is often more than one way of creating the same object. Sometimes in 3D Studio it is easier to create objects in the 3D Editor instead of importing them from the Lofter, i.e. a sphere. The Editor offers commands to create a Box, Sphere, Cylinder, Tube, Torus and Cone.

The Array tool copies an object many times in either a linear or radial manner. This tool is particularly useful for creating scenes such as fences or when a number of objects are to be arranged in a circle.

Boolean operations in the 3D Editor allow overlapping objects to be joined, subtracted or intersected. The Boolean command can be used to pierce holes in diverse shapes or to create complex intersections. It can also, for example, 'drill' a hole in an object which does not go all the way through. Boolean operations however, do not always work the way you think they should, particularly on complex shapes. A Hold should always be applied when using Boolean operations as the geometry of the original objects is destroyed.

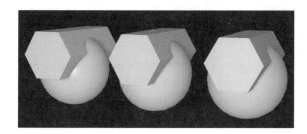

3D Boolean

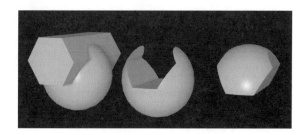

11.4.7 COMBINING, COMPOSING AND ALTERING 3D FORMS

Within the 3D Editor there are a huge number of ways to manipulate the 3D geometry, both to further refine individual shapes and to build more complex objects and scenes. By using Selection Sets, Objects and Elements can have parts of their geometry selectively altered. Single or groups of Vertices and Faces can be altered using commands such as Move, Rotate, Scale, Skew, Mirror, Bend and Taper. As you can imagine, once you get down to being able to manipulate every Vertex individually, the permutations become infinite. However as a rule, the majority of alterations are done to Selection Sets of Vertices and Faces and to whole meshes. Some commands only work on sets of Vertices and Faces. You cannot for example Bend or Taper a single Vertex, it can only be tapered or bent in relation to other Vertices.

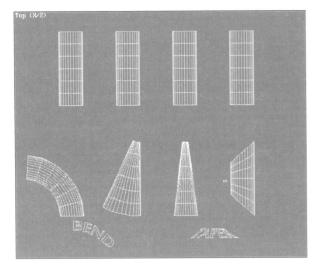

The position of the Global Axis and the use of the Local Axis can be important factors in these commands. The Tab key is also used as a toggle to define whether objects can be moved in any direction or to restrict their movement to the horizontal or vertical. Snap and Angle Snap are constantly used features in the movement and manipulation of objects in the 3D Editor.

Bend and Taper and the affect of the axis position

Groups of objects can also be combined. Often when working with a scene, it is easier to combine several objects into a larger object. This is done once again to rationalise the display and to make any mistake when moving altering objects, less likely. Several objects can be joined, they adopt the name of the last object picked and become Elements of that object. As the 3D Editor has the same manipulation commands for Elements as objects, this does not usually create any modelling problems. The example used before was the one of the car. The wheels may be created and joined to the body as Elements. This way if you want to scale the car, there is no risk of forgetting to select and scale the wheels also.

11.4.8 OTHER MODELS AND METHODS

Once a user becomes competent with the basic 3D creation and manipulation commands, there continues to be an enormous number of techniques and skills which enable almost anything that can be imagined to be created. External 'third party plug-ins' such as IPAS routines can be used for procedures like automatically modelling complex trees.

For the average user, it is not usually necessary to use such advanced tools, as there are already such a huge number of ready made meshes available for all the versions of the program. The World Creating Tool kit which is supplied with 3D Studio for example, has dozens of meshes in it which can be incorporated into scenes. Others can be purchased from third parties and sometimes they come free with specialist books and magazines. Meshes can also be found, and downloaded, from user group sites on the Internet.

Externally sourced meshes are really worth exploring and building into a library. For example if 'any old house' is needed in the background of a scene, it is much quicker to Merge an existing mesh than create a new one. Also, because these meshes are usually made by experienced users, they have been optimally produced and contain relatively few vertices, often many fewer than a beginner would use.

Flat shapes (shapes without any thickness) can be brought directly into the 3D Editor from the Shaper. This is particularly useful for lettering to be used as decals, as once imported, these objects are flat faces and contain few vertices. On the following page the 3D figure shows some imported flat text. The vertical 3D is the same text which has been imported, via the Lofter, where it has been given some thickness.

*The 3D image
demonstrates
imported flat text.*

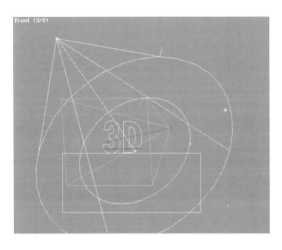

To complete a scene, it needs to have lights, objects need to be given surface qualities and cameras need to be placed to view the scene. The 3D Editor world has no illumination. Adding lights enable the objects to be viewed, have mood and effects added and to bring focus to a specific area in a scene. All newly created models are assigned the default material which is a matt white. The 3D Editor offers a variety of ways of giving meshes surface qualities which, in many cases, simulate materials in the real world and finally a camera creates a perspective view of the object or scene.

Cameras and Lignts

11.4.9 LIGHTS

These 'virtual' lights of 3D Studio work differently to real lights. For example, on the standard setting, 3D Studio lights illuminate equally from any distance. Also (with the exception of spotlights set to cast shadows) lights will illuminate through objects and will equally effect an object which is behind another in the foreground.

Lights in 3D Studio do not reflect or radiate from surfaces in the scene. Objects which have shiny, reflective or glowing finishes, which you would expect to affect neighbouring objects, do not. This effect, called Radiosity, can be handled by some rendering packages, but not by 3D Studio.

3D Studio's 3D Editor cannot deal with refraction (the distortion which occurs when light passes through transparent materials of different densities). Neither can coloured shadows be cast from translucent, coloured objects. Most of these effects can be simulated by the skilful use of self-illuminating materials and the positioning of coloured lights and do usually not restrict creativity. MAX and VIZ however can simulate these effects.

There are three types of light used in the 3D Editor - Ambient, Omni and Spot. Ambient is the background light which illuminates everything in the scene. It has no directional component and is used to determine the general brightness and colour in the scene. Colour can be assigned to a light using the Red Green Blue (RGB) or the Hue Luminance Saturation (HLS) sliders, which appear in a dialog box when the command Lights/Create is given.

Omni lights illuminate in a spherical pattern. They too can be varied in brightness and colour and cast a broad omnidirectional light within the scene. They only illuminate the sides of the object facing the omni light and as they do not cast shadows, their light is not blocked by objects in the scene. They can be adjusted in a number of ways through the Omni Light Definition dialog box.

Spotlights cast light as if projected from a point in a conical fashion. They can be varied in brightness and colour and have the same features of Omni lights of Multiplier, Attenuation and Exclusions. They have two radiating cones, a 'hot spot' and a 'fall off'. Each of these can be independently varied in size, up to an angle of 175°, to alter the intensity and 'shape' of the light across its field of

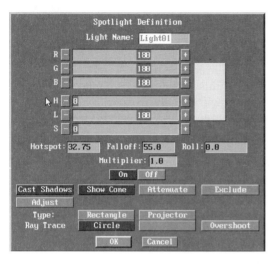

Spot Light Definition dialog box

projection. They can be made to cast a circular or rectangular beam of light, can be used as a projector by shining them through an image, like a slide projector and, most useful of all, they can cast shadows.

The Multiplier sets the strength of the omni and spotlight, this can even be set to a negative value where the light becomes a light 'vacuum' removing light from a scene. Attenuation allows the power of the Omni light to be set to diminish over distance. Exclude brings up a list of objects in the scene, by clicking on an object name, you can choose not to have that light affecting that object.

Shadows are either Mapped or Ray traced. Shadow maps work by creating an approximate image from the perspective of the shadow-casting spotlight. The sharpness of this can be adjusted with Shadow/Parameters. This is the type of shadow which is used, for example, with indoor lighting, where shadows have softer edges.

Ray traced shadows, as the name suggests, trace the spotlight rays through the geometry. This type of Spotlight takes transparency, but not colour, into consideration. It casts a crisp, hard-edged shadow. This can be useful for simulating sharp outdoor shadows, like those cast on a sunny day.

Spotlights are the most used type of light to create drama and realism in a scene. However shadow-casting spotlights can slow up rendering and significantly increase file size so they should be used sparingly. Two or three shadow-casting spotlights are usually enough in any scene.

11.4.10 ADDING MATERIALS IN THE 3D EDITOR

Which materials are chosen and how they are applied to objects, is one of the most important and skilled parts of 3D modelling. This is how 'reality' is simulated in the computer, or at least the creator's own version of 'reality'. A material, as defined by 3D Studio, can be more than just a coat of coloured paint. Every aspect of the way a surface appears can be defined. Surfaces can have qualities added to them which makes them shiny, matt, bumpy, reflective, metallic, transparent or any combination of these. Created in the Materials Editor, they can then be applied to objects, elements or individual faces.

11.4.11 APPLYING MATERIALS

Materials can be applied to the geometry in the 3D Editor, in a manner similar to applying wallpaper to a real life object. 3D Studio 'wallpaper' is a little more flexible than the sort you attempt to paste at home, but it is a good analogy.

To apply a basic colour to a whole object, a colour is chosen and assigned to the object. That whole object will then become, for example, a gloss red, which simulates red plastic. If however there is more information in the materials surface, i.e. it is bumpy or has a wood grain or pattern, then where and how that surface is applied needs to be defined. This is done in 3D Studio, as it is with most 3D programs, by 'mapping' this information on the surface.

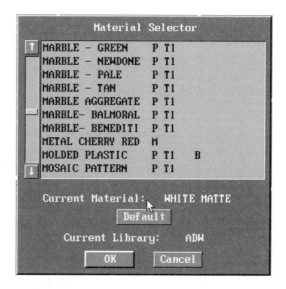

Select Materials
dialog box

11.4.12 MAPPING

Mapping defines how the 'wallpaper' is applied by placing a Bitmap onto the surface of the geometry which acts like a framework to 'hold' the image. There are two methods of placing maps on the surface. Tiled mapping is repeated seamlessly across the surface of an object and decal mapping is used to repeat an image, with spaces between, across a surface like a polka dot plate. Decal mapping also can be used once as a label or 'decal' on an object such as a bottle.

There are also several different types of map. Texture Maps apply a bitmap or material to the surface of the mesh. This is actually an image or 'photograph' of a material which is 'painted' onto the object. 3D Studio supports several common file formats for texture maps including - GIF, Targa, Tiff, JPEG and Cel.

Opacity Map uses the intensity of the colours in an image to create varying degrees of transparency in the bitmap. This can be used for example, to create the impression of a complex window where the detail or leading is defined as the opaque part of an opacity map.

Bump Mapping takes the image assigned to it and at rendering, gives the impression of a 3D or embossed surface. It does this by interpreting the intensity of colour in the image as the heights of bumps off the surface. The lighter the intensity, the higher the 'bump'.

Shininess and Specular mapping are similar to each other and modify the highlights which are used to give the impression of shininess on a material.

Self Illuminating mapping is used to make it appear that a material produces its own light. This is useful, because as mentioned in the section on lighting, 3D Studio does not deal

Tiled and Decal mapping

with radiosity and using self illumination can be used to simulate that phenomenon.

Reflection Mapping gives objects a reflective quality. There are four options - Spherical, Cubic, Automatic and Flat. Reflection Mapping works well on flat surfaces. It can be used in combination with another texture map, marble for example, to give the impression of a high gloss surface. Because 3D Studio does not have a Ray tracing render option, creating complex objects with reflective surfaces is a fairly complicated procedure. It is best simulated using what is called a Cubic Environment Map. This takes six views of the scene, as if from inside the object, creating six Targa bitmaps. These are then assigned to the surface as bitmaps and give the impression of the scene being reflected in the object. The flower vase in the scene on the back cover uses a Cubic Environment Map to give the impression of a shinny metal surface reflecting the other objects in the scene.

Checker opacity map applied to a sphere

11.4.13 MAPPING COORDINATES

How these different types of map are applied is extremely important. To simulate successfully a piece of wood for example, the grain of the wood needs to be of an appropriate scale and pointing in the right direction. 3D Studio offers several different mapping types:-

~Lofted mapping is 'built' onto the object in the 3D Lofter.
~Planar mapping projects the bit map perpendicularly onto the surface as if from a flat plane, and then repeats, or tiles it, until the surfaces are covered.
~Spherical works as if the map is projected inwards form a sphere.
~Cylindrical is projected as if from a cylinder.

All the mapping types can be scaled, moved and altered in a variety of ways until their scale and orientation is appropriate to the object or scene.

It is often better to use a texture map on the surface of a simple object than to model a complex mesh. Take a distant building in a scene for example. Instead of modelling all the doors and windows as meshes, which would be time and memory intensive, images of a building can be mapped onto the simple box.

Mastering the use of mapping coordinates and understanding how they affect the way materials are applied to objects, is vital for good modelling. Different modelling packages use different conventions. 3D Studio's method for applying mapping coordinates initially appears complex but is quickly mastered.

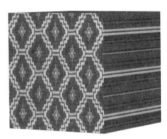

Planar mapping   *Spherical mapping*   *Cylindrical mapping*

11.4.14 CAMERAS

Cameras are a way of viewing the scene in perspective. They can be manipulated in several ways and use 'real' camera conventions. You can, for example, change the lens focal length and dolly the camera in and out. Although a camera is not needed to render a scene (scenes can be rendered in orthographic or flat view) it is the perspective of the camera views, and the ability to understand and manipulate these views, that enable reality to be simulated.

The cameras in 3D Studio stop short of real cameras in that they do not have a depth of field (i.e. the 'depth' of the image which is in focus in the scene which, with a sophisticated camera, can be controlled). Once again there are ways that 3D Studio can simulate this blurring. It uses Fog and Distance Cueing which enables the colours of objects to be 'faded out' according to the distance from the camera, or for fog to be added in the same way.

Stock Lenses (28mm, 50mm, 135mm, 200mm etc.) are offered for easy selection when the camera is being created, however they can be further refined to achieve the best view of the object or scene. It is useful to remember that the human eye sees approximately the same as a 50mm lens. Below this focal length you tend to get an exaggerated perspective, and above it the perspective tends to diminish. The position, perspective and field of view can be used to dramatically effect the nature of the image. A low viewpoint with exaggerated perspective, can create the impression of a object being 'large and looming' while a distant camera viewpoint and a long lens length can make objects 'distant and insignificant'.

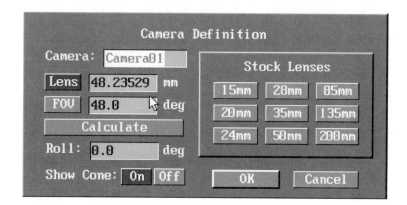

Camera definition dialog box

11.4.15 PREVIEWING THE IMAGE

Release 4 offers a real-time preview facility in the 3D Editor. An
active camera viewport can be rendered in basic way using Phong,
Gouraud, Flat or Wire shading. By moving the mouse in the
viewport, 'real-time' movement of the objects in the scene is possible.
This rendered preview can be used to adjust the camera settings and
once a desired view is achieved, the cameras settings can be saved.

   This makes it possible to 'match' the perspective of the created
object against an image of a background. This is useful, for example,
for placing the model of a new building in the image of an existing
site.

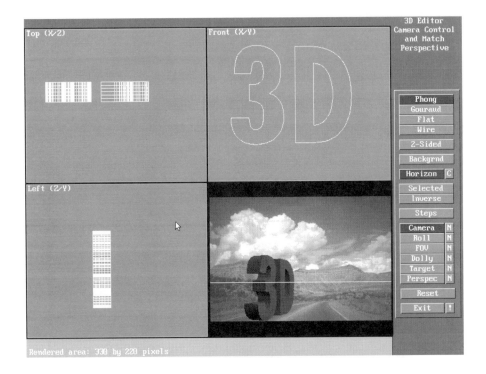

*The 3D Editor's real time
preview facility (F7) used to
create a perspective match*

11.4.16 RENDERING THE IMAGE

Once the objects have been created, given their material qualities and had lighting added, the scene (or a portion of it) can be rendered in the current viewport in the 3D Editor.

The Render Still Image dialog box allows all the rendering parameters to be adjusted. For the majority of still rendering, the default is the best one. However there are sometimes changes that need to be made to the parameters.

Flat, Gouraud, Phong or Metal are offered as rendering shading models. Phong and Metal are the most realistic shading levels, but of course the slowest. To see shadows, bump maps and reflection maps either Phong or Metal must be used. For test renderings, to check for example the lighting in a scene, the renderer is often set on a quicker mode (such as Gouraud) to save time.

Anti-aliasing, Filter Maps, Shadows, Mapping and Auto-Reflect are usually left on for final renderings. The main reason for turning any of them off would be to speed up rendering during a test render. It is sometimes necessary to turn Force 2-Sided on if rendering the inside, as well as the outside of an object. If during rendering any objects appear to have 'transparent' parts in them, turning Force 2-Sided on usually solves the problem. The Force Wire is useful as it renders all the objects as wire frames.

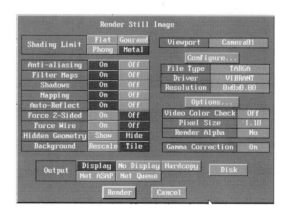

Render Still Image dialog box

The file type and resolution are set up in the Renderer Configuration dialog box which is accessed through the Render Still Image dialog box. The file types that 3D Studio renders to are Gif, Targa, Tiff (colour and mono), BMP and JPEG. These are all common file formats and are interchangeable with many other programs on a range of platforms. The resolution of a rendering within 3D Studio is defined as width and height of pixels (i.e. the standard VGA screen the setting is 320x200 pixels). Better quality displays and more advanced graphics cards can enable much higher resolution images to be viewed. By choosing the 'Null' display parameter, an image of any size can be rendered (i.e. not restricted by the screen output resolution). The image can then be viewed in a 2D program, such as Adobe Photoshop.

Using a Null display and rendering an image with a high pixel width and height is recommended if the still image is intended for print. It has to be remembered of course that the higher the resolution, the bigger the file size.

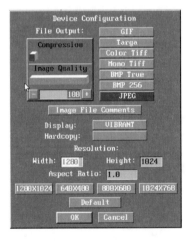

Renderer Configuration
dialog box

11.5 THE KEYFRAMER

The Keyframer is where animations can be produced with the objects
and scenes that are created and assembled in the 3D Editor. This is a
very powerful part of the program, and combined with its highly
flexible modellers, it has made 3D Studio the choice of many professional animators and computer game developers.

11.5.1 DIRECTING THE ACTION: ANIMATION

When creating an animation you are 'directing the show'. All aspects
of the objects and the scene can be controlled, and referenced against
time. With computer animation, unlike film animation, the rate at
which animations can be played back, is determined by hardware
and software limitations. The internal animator within 3D Studio
runs at about fifteen frames per second (15 fps) which achieves
acceptable smoothness.

   The interface looks almost the same as that of 3D Editor. There
are the same viewports, the same pull-down menus and a similar
icon panel. There are also a new set of commands which relate
purely to animation. These are Hierarchy, Object, Lights, Cameras,
Paths, Preview, Renderer, Display and Time.

   TRACKinfo and KEYinfo are dialog boxes which are the main
'directors tools' for setting up the animation. There are buttons
which let you determine which segment of the animation is active
and being worked on. (Animations are usually made in segments
and these are pieced together for the final show.) You can jump to
the end or beginning of an animation, move one frame at a time or
'play' the animation.

   When the mouse is moved to the bottom of the display, a Time
Line appears. The number in this line is the current frame number
and by grabbing the number with the mouse and pulling it back and
forward, you can move through the animation frame by frame.

11.5.2 KEYS AND KEYFRAMES

3D Studio uses the common computer animation convention of having Keyframes (hence the animation module's name). These Keyframes define the point of any change in the actions or attributes of objects, lights or cameras, during the animation. At frame zero, all the start positions and attributes of each individual component of the scene are recorded with Keys. If any of the positions or attributes of the components change at any point in the animation, a new Keyframe is created to 'record' or 'mark' that change. The computer system then calculates what would have happened between these two points, and records these interim stages as frames. So, Keys define changes at a component level and Keyframes at a frame level

The most basic example animates constant movement of an object over a specific time. For example a ball moving from one point to another in one second at 30fps, the frame is set to zero and the ball placed in its start position. Then the current frame is set to thirty, and the ball moved to its end position. The software will calculate the interim positions and the ball will move with a smooth constant motion between the two points.

Placing new Keyframes in the animation is quite straightforward. If the frame number is changed and some movement or alteration is made to anything in the scene, this creates new Keys at this Keyframe. Take the moving ball as an example. If as well as making it move smoothly to frame thirty, it 'squashes' in the middle of the animation, frame fifteen is made a Keyframe. By using either the 'Go-To Frame' dialog box or pulling the number along the Time Line, frame fifteen is made current and the ball is 'squashed'. On playing the animation, the ball will move off, as it is moving it will squash in the middle, and then return to its original shape at the end of the animation segment, all in a constant motion.

Keys can be attributed to lights and cameras as well as objects like the ball. They can move with, or separately to, objects and with a Spotlight or a Camera, you can define and animate both the source and the target points. Any adjustments that you can make to either a light (i.e. colour and brightness), or a camera (i.e. field of view or roll), can be 'Keyed'.

11.5.3 PATHS

If the position of objects are changed during the animation, a Path is
generated by the system between the Keys of any moving object.
This is the route it will follow at animation. The line that results is a
'splined' line in either 2D or 3D. Motion paths are shown as red lines
on the screen and white squares represent the individual frames. The
spline line that 3D Studio puts between each of the key points,
because of its averaging, is not always the motion that is expected or
desired. The motion can be more precisely defined by importing a
Path from the 2D Shaper or the Lofter or from another program
(such as a 2D or 3D Polyline from Autocad). If you define the path in
this way, it is called Path Based Animation. Each vertex of the
imported path is treated, and can be adjusted, as a Keyframe in the
animation. The figures below show a basic spline path that the
Keyframer has drawn between three specified points and a Helix
path which has been imported into the Keyframer from the
3D Lofter. Other aspects of the movement along the animated path

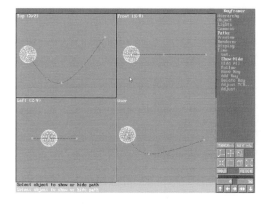

A basic spline path created by the
Keyframer between three specified points

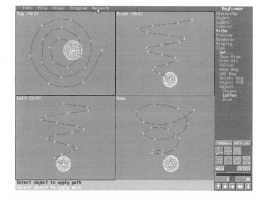

A helix path which has been
imported into the Keyframer

can be controlled. Path Follow for example, controls whether the object following the path will 'bank' around the corners and if so, at what angle. But the real control over what happens at each individual key, is controlled by the Keyinfo dialog box.

11.5.4 THE KEY INFORMATION DIALOG BOX

The Key Information dialog box is where adjustments to the key values of an object, light or camera are made at any established Keyframe. The values for a mesh object, it's Position, Rotation and Scale at that point, appear and can be edited. For an Omni light you can adjust Position and Colour; for a Spotlight Position, Colour, Hotspot, Falloff and Roll; and for Ambient light, the Colour values. With a camera you get the opportunity to set Position, FOV and Roll. In fact you can alter, through the Keyinfo dialog box, almost all the basic settings of an object, light or camera. But in reality, the material in this dialog box is really for information, and most of the settings are much more intuitively adjusted in the Editor or through the main Keyframer interface.

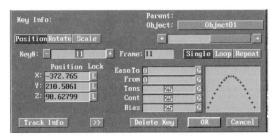

The Key Information dialog box

In addition to these basic settings, there are also a sophisticated series of adjustments that can be made which influence the way an object moves into and leaves a Keyframe. Ease To and Ease From, influence the velocity of a movement into, and out of, the Keyframe. When Ease To is set to high, the motion slows down as it approaches the Keyframe, and when Ease From is set to high, the motion is slower when it leaves the Keyframe. These settings would be used, for example, if you are trying to give the impression of a object bouncing on the end of an elastic band. As the object on the elastic moves toward the bottom of is motion, it will slow down almost to a stand still and then accelerate off back up under the tension of the elastic. In this example the Ease To will be set high, and the Ease From set low. **Tension** controls the overall curvature of the 'spline", **Continuity** affects the angle at which the spline enters and leaves the Keyframe and **Bias** affects the overshoot or undershoot of the motion through the Keyframe.

These are all combined graphically in a window in the Key Info dialog box, which draws a graph of black Xs which attempts to illustrate the nature of the motion. The figures below illustrate how TCB settings affect various simple motion paths. (The centre one shows the default settings.) These commands bring greater realism to the motion of objects in an animation.

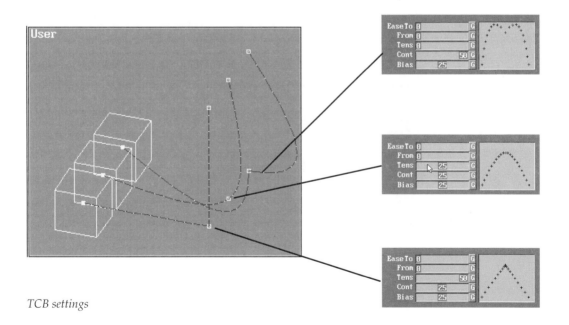

TCB settings

11.5.5 THE TRACK INFORMATION DIALOG BOX

The main way of seeing, and controlling, what happens along the whole path of an animation, is through the Track Information dialog box. While the Keyinfo dialog box adjusts key settings for individual components in each Keyframe, the Trackinfo dialog box allows you to view the Keyframes in relation to each other and to the rest of the animation. This is graphically displayed, as it is in many animation programs, as a graph with the number of frames along the horizontal axis and the different Tracks along the vertical. What Tracks are shown in the dialog box depends on whether a object, light or camera is chosen. All the information in each Track (i.e. with a mesh object they will be Position, Rotate, Scale, Morph and Hide) is displayed as a vertical row of black dots. At frame zero, most Tracks will have a dot in it depicting the status of the object, light or camera at the beginning of the animation. If there is no change in a component part's position or attributes during the animation, there will only be a row of dots at frame zero and then again at the final frame. If there is any change, another dot will be placed in that Track, at each Keyframe, to mark that change.

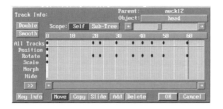

Trackinfo dialog box for the head of the 3D Studio sample animation 'Bird Walk'

It is by working between the Track and Key Information dialog boxes that the 'action' is directed.

11.5.6 MORPHING

In 3D Studio, morphing is handled in the Keyframer and there is a track in the Trackinfo dialog box which controls morphing of objects in an animation. There are some restrictions, each object to be morphing from or to must have the same number of vertices, or 3D Studio cannot morph them. The best way to ensure that two objects can be morphed is to create them in a similar way in the Lofter. For example two 2D Polygons, both with the same number of vertices but different shapes, can be sent along Paths which are different but have the same number of steps. Another alternative is to copy and alter objects in the 3D Editor.

11.5.7 HIERARCHICAL LINKING AND INVERSE KINEMATICS (IK)

These are fairly complex procedures which 3D Studio uses to introduce greater realism into character animation such as of a person or animal. In Hierarchical Linking, various parts of the character are linked together using the 'parent child' analogy seen in chapter 4. The figure on the next page shows part of the Hierarchical link information in the Bird Walk animation.

Inverse Kinematics (IK), available in Release 4, develops Hierarchical linking and allows 'end-effector oriented' animation. For example if the hand is moved, it will effect the forearm, it in turn the upper arm, and so on. IK assigns a 'skeleton' to the character. Physical constraints can be assigned to the joints in this skeleton to limit their range of movement. Taking the knee joint as an example, it can only bend in one plane and the angle of this bend is limited to approximately 160° from straight to fully bent. The wrist joint is much more flexible and has a cone shaped area of movement. These types of limitations can be applied to joints in the models using IK to give greater control and realism to the characters and create lifelike motion. The figure below shows the skeleton the IK process has created for the Bird Walk animation.

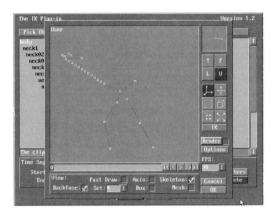

The skeleton the IK process has created for the Bird Walk animation

11.5.8 PREVIEWING AN ANIMATION

As an animation is a large number of still images, rendering these images can take a long time, particularly if they are being rendered to a high resolution. As a result, before any animation is set to render, there are several methods of previewing it. Previews are constantly made in the often slow and painstaking world of computer animation. Often a single move, of a single object, is modified over and over until the motion is correct. Then that motion is previewed within the scene, to see how it relates to motion of other objects, and so it goes on.

There are a number of previews offered in the Keyframer. Box Mode displays the objects as bounding boxes that enclose the outer extremities of each object, and is the quickest to redraw. It is fine for checking basic movements in the animation. Fast Draw, which is a new feature of Release 4, works very like the still image Camera Preview in the 3D Editor. This creates a simple preview in the camera viewport, with one basic light, surface colour but no materials or mapping. It will produce the preview very quickly in either Wireframe, Flat, Gourand or Phong. Phong is the default.

These basic previews are fine for checking motion, however to test the other aspects of the animation before final rendering, experienced animators usually test animate every 'Nth' frame. By rendering every, say, fifth frame, a good impression is gained of how the

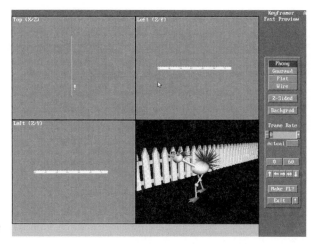

Animation preview

final product will look. If there are errors, it's back to the Keyframer to fix them and repeat the process.

11.5.9 RENDERING AN ANIMATION

Rendering an animation is the same as rendering a still frame, except of course multiple frames are being rendered. All the image and file output variables in the Keyframer's Render dialog box are the same as those in the Still Image Render Dialog box in the 3D Editor. Animations are saved as FLC files. They can be stored and viewed on the computer or output via appropriate hardware to video. The number of frames per second produced in an animation should always reflect the final output, i.e. 24 fps, and not the 'playback' capabilities of the PC which is currently unlikely to exceed 15fps.

Rendering an animation can take a long time since each frame can take many minutes to render and there may be hundreds of frames. This highlights how important test renderings are and why animations are often programmed over night to render. Professional animators or game developers who are using 3D Studio for animation often use either powerful workstations, such as Silicon

179

Graphics, or network together a series of PCs to render animations. All the modelling and still image work that you can do with 3D Studio is within the capabilities of most decent newer PCs. However once you get into animation, you can quickly reach their limits!

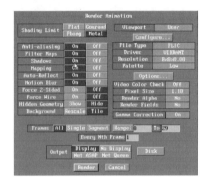

Render Animation dialog box

11.5.10 THE VIDEO POST

The Video Post can create special effects that cannot be handled by other parts of the program. It can Image Process, or alter, every image in the animation, after it has been rendered but before it is saved. This is done with what are called IPAS routines which are usually algorithm-based and are 'plug-ins' into 3D Studio. They can produce the same type of effects which are available for still image processing in programs such as Photoshop. A common IPAS routine is 'ripple' which can be used to give the effect of the whole animation taking place underwater.

It can combine two or more animations into one, can apply travelling and static 'mattes' and create visual effects such as 'transitions' and 'fades' between animations. It also contains the Alpha Channel which enables a static image to be composed with an animation. For example, a rendering of a proposed building could be superimposed on to a digital photograph of the site. The Video Post is quit complex and is really the preserve of the experienced animator and not the beginner.

11.6 THE MATERIALS EDITOR

The Materials Editor in 3D Studio is a very powerful and versatile surface creation 'laboratory'. It is where materials are created, adjusted and then saved to be applied to meshes in the 3D Editor. Once again there is a myriad of parameters which can be set to influence how a particular material appears at rendering. While all the other modules 'pass' geometry from one to the other (and hence have a lot of common features) the Materials Editor is like a stand-alone part of the program, only 'passing' material information back into the 3D Editor. The Materials Editor has a very different interface to the other 3D Studio modules (it is however similar to the Materials Editor in MAX and VIZ).

11.6.1 HOW 3D STUDIO MAKES A MATERIAL

Materials are created by having various attributes such as colour, shininess and transparency assigned to them. This information is stored, and any bitmap information is added. The two things together define the material. The material however, as it is named and stored on file, does not contain all the bitmaps that it uses to define itself, it simply 'tells' the renderer what files to use, and where to find them. Therefore bitmaps files may be used to create a number of different effects, on a number of different materials, whilst only existing once on disk. (For example, the file Sand.CEL is used as a Texture Map in the 3D Studio material Sand, and as Bump Map in the material Bumpy Red Plastic). All the files that are used for bitmaps, are usually kept in the one MAPS sub-directory (although this can be changed and added to). This is where the program looks for these files at rendering and, if they are not there (or in another defined Map Path as it is called) the rendering will fail.

So a material is defined as having some 'attributes' and some 'pointers', and how it looks rendered on a particular object, is determined by the mapping coordinates that have been applied to that object. As with meshes, there are a huge number of materials which are freely available to be used as bitmaps. This saves a lot of time, as not only can they be imported into the program (with the appropriate files being put in the appropriate Map Paths) but these image files can be used as a basis for new materials. The World Creating Tool kit shipped with 3D Studio R4, has a good selection of materials which can be added to a personalised library. For example, if a scene demands a particular type of wood such as Maple or Oak, these and many others are available as proper tiled bitmap images on the CD which can be imported into the Materials Editor, and applied to meshes.

11.6.2 THE INTERFACE

This module's interface, similar to the Materials Editor in MAX and VIZ has the pull-downs Info, Library, Material, Options and Program. A series of windows are used for rendering samples. Up to seven different samples can be displayed at once, but only the one that is highlighted and in use is displayed in colour. A sample of the material is rendered on a sphere or cube. The default background in the sample square is black although it can be made a multi-coloured checked pattern which is very useful when you wish to check an object's transparency. In the middle of the figure to the left, a Frosted Perspex material is placed in front of the checked pattern and gives a good indication of its translucency.

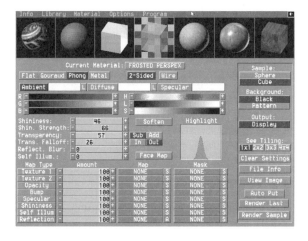

Materials Editor interface

11.6.3 ATTRIBUTES

The Materials Editor defines an object's colour is three different ways - Ambient, Diffuse and Specular. The Ambient colour defines the shade of the object's colour, when it is not directly lit (i.e. areas in shadow). The Diffuse colour is the main solid colour of the material and is the most important setting. The Specular setting affects the colour of highlights which appear when light sources cause highlights on shiny surfaces.

The colour of the material is determined in the same way as it is with lights, by the settings in the Colour Sliders. There are two sets of these; one set defines the colour in terms of the amount of Red, Green and Blue (RGB) and the other in terms of Hue, Luminance and Saturation (HLS). Both are interrelated and reciprocal, and can be used separately or together to define and 'fine tune' a colour. Using a paint mixing analogy, you might mix the colour with the RGB sliders, and then modify it by adding white and grey using the Luminance and Saturation sliders. Move one set of sliders, and the others move automatically to their corresponding values.

Another set of sliders adjust Shininess and Shin Strength which affect firstly the size of the specular highlight (the higher the value of Shininess the smaller the highlight), and secondly the intensity of it (the higher the Shin Strength, the closer the light appears to move towards the material).

The Transparency slider determines how transparent an object looks. Set at 0 the object is opaque and at 100 the object is completely transparent. Unlike MAX and VIZ, 3D Studio does not deal with refraction in transparent materials. Self illumination (to make it appear glowing) is a useful attribute to be able to give a material as it can be used for objects like light bulbs, combined with a texture map to create effects like a TV screen and, to compensate for the fact that the renderer does not deal with radiosity, to give the impression as though light is 'bouncing' from another material.

11.6.4 MAPS

The other main sliders define what, if any, bitmaps are used in the material and the percentage of that bitmap. Texture, Opacity, Bump, Specular, Shininess, Self Illumination and Reflection can applied in a variety of combinations and percentages.

The figures below demonstrate this. In the figure on the left, the material White Bumpy Plastic has various amounts of the Bump Map Sand applied to it. The four samples have, from the left, 0%, 10%, 50% and 100% of the Bump Map applied and its effect is obvious. In the figure on the right, a combination of different types of bitmaps are used to make the material Corroded Metal. It has 100% of the JPEG Oldmetal as a Texture Map, 29% of the same JPEG as a bump map and 47% of the Targa file Valley_L, as a reflection map. This demonstrates how more complex materials are made.

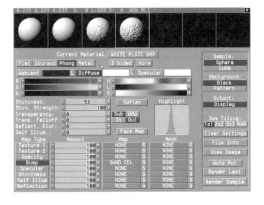

*The material White
Bumpy Plastic*

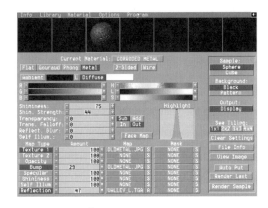

*The material
Corroded Metal*

184

11.6.5 MATERIAL LIBRARIES

Materials in all versions of 3D Studio are stored in libraries. When you load the program, a default library (3DS.MLI, MAX.MLI or VIZ.MLI) is loaded automatically. These are very good, large libraries of pre-prepared materials that are supplied with the programs. Since each material contains parameter values and is really only 'pointing' to the texture map files it uses, material libraries take up very little disk space. As a result, most users begin to build up their own materials libraries, creating new libraries of materials based on different permutations of the bitmaps which already exist in the Maps sub-directory, incrementally adding new bitmaps and developing new materials.

CHAPTER 12

3D STUDIO MAX

3D Studio MAX is another more recently developed version of the program. It has some different, more powerful modelling and animation features, and is used extensively by 3D game developers and film animators.

12.1 HARDWARE REQUIREMENTS

The big difference between the DOS versions of 3D Studio and MAX, is that MAX runs on the Windows NT operating system. (It can also run, but not as efficiently, on Windows 95.) MAX demands a much greater hardware specification than 3D Studio for DOS, and this can put it out of reach for some users. The processor should be Pentium 90 or better. MAX will benefit from the fastest processors available (300MHz+ at time of writing.) Multi processor machines can be exploited because of the NT platform's ability to multi thread. 32 MB of RAM is an absolute minimum, 64-128MB is more realistic for complex work. There needs to be 200-300MB of hard disk space available to be used as swap space. MAX is designed to be viewed at 1024x768 resolution at 24-bit colour. It is best therefore to have a good graphics card with at least 4MB of RAM and as large a screen as possible.

12.2 THE INTERFACE

MAX has a very different user interface from Dos 3D Studio. As it runs under Windows NT (or Windows 95), MAX makes much greater use of icons and dialog boxes to give commands and change settings. It does not have the five different modules, but one interface which handles all the tasks. This makes it packed with icons and necessitates the large screen size.

The default screen layout has four viewports, top, left, front and perspective, (not unlike the 3D Editor in 3D Studio). This can be changed to any configuration through the Viewport Configuration dialog box (this is accessed through the Views pull-down menu or by clicking over the title of the Viewport). Also in the Viewport Configuration dialog box, how the objects are displayed can be set from a simple bounding box, through wire frame to smooth rendered. Unlike 3D Studio for Dos, MAX can display interactive rendered viewports which allow real-time viewing of the scene in a

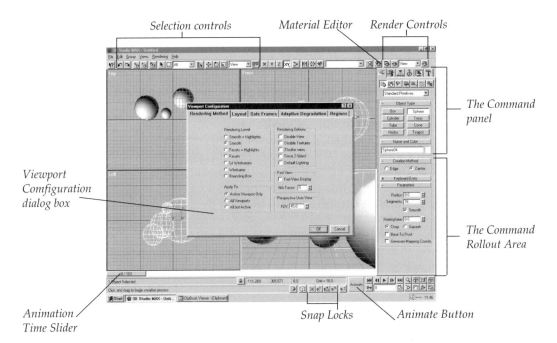

The MAX interface

rendered form. This gives an instant visual feedback to all additions and modifications to the scene or animation. However, the higher the quality of the shading, the slower the refresh. (This can reduce to a crawl if the machine has not got 64+MB of RAM). Buttons on the lower right of the screen allow several panning and zooming options in the active viewport.

12.3 ACCESSING COMMANDS

The MAX interface provides a variety of methods for accessing commands. Pull-down menus, located at the top of the display, are File, Edit, Group, Views, Rendering, and Help. Icons, logically grouped in Toolbars along the top and bottom of the interface, are the user's access to the most commonly used commands. When the mouse is held over an icon, a 'tool tip' appears describing the command. Some icons contain sub-sets of icons which appear when the mouse button is pushed and held.

The Command Panel is a significant feature of the MAX interface. On the right of the display, it gives access to all the drawing and editing commands. Each is made up of three areas. At the top, there are buttons for Create, Modify, Hierarchy, Motion, Display and Utility. When pushed, they display a sub-set of commands, and below each of these there are 'Roll-outs' where the parameters are found.

While the interface and the methods of giving commands in MAX are completely different than the DOS based versions of 3D Studio, many of the ways that the program deals with the creation and manipulation of objects and scenes, are very similar.

Create Command Panel

Modify

Motion

Display

12.4 CREATING OBJECTS

The commands for creating 3D objects are in the Create command panel. MAX defines these in five categories:

Standard Primitives, the usual objects such as box, cone, sphere, torus and so on. These are the basic building blocks of much 3D work. In Create you can also find Hedra (which makes a geometric shape such as Tetrahedron or Octahedron) and Teapot (based on the University of Utah's model which in the early days of computing represented a complex modelling task).

Create Hedra dialog box

Create Teapot dialog box

Teapot
standard primitive

Hedra
standard primitives

Patch Grids, are an exciting new feature of MAX. This command creates a Bézier surface, a surface of squares or triangles, that can be adjusted through the manipulation of control points. When a control point (or vertex) is moved or adjusted, it can be made to affect an area of the Patch Grid and not just the single vertex. With this you can pull vertices and create non-uniform curved surfaces.

'Pulling' a form using a Patch Grid

Compound Objects, combine two or more objects into one. The most common form of which is the familiar Boolean operations of Intersection, Union and Subtraction.

Loft Objects in MAX, work almost identically as the process explained in the earlier section on the 3D Studio Lofter. 2D lines, or shapes, are created in the Shapes Command Panel. (In 3D Studio, this is done in the 2D Shaper.) One line or shape is chosen to act as the Path, and one or more closed Shapes are chosen to be Lofted. The same restrictions for Paths and Shapes apply as in 3D Studio for Dos.

As was explored in the section on 3D Studio, the real power of the lofted (or swept) object is the ability to deform the object's cross section and angle as it travels along the path. MAX has the same Deformation Tools as 3D Studio - Scale, Twist, Teeter, Bevel and Fit. The Deformation Tools are controlled by very similar Deformation Grids as those explained in the section on 3D Studio's Lofter. Unlike the Deformation tools in the Lofter in 3D Studio (which are applied while the model is being lofted) the deformation tools in MAX are applied as Modifiers. Selected from the Modify menu, a dialog box appears in the main screen which displays the deformation curve to control how it affects the object. Extrude and Lathe are terms used in

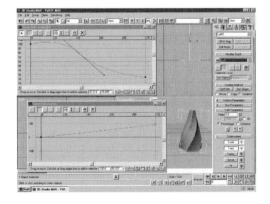

The Scale and Twist Deformation Tools

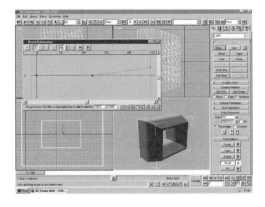

The Bevel Deformation Tool

MAX and not in 3D Studio. Extrude gives depth to a 2D shape in the same way as a default straight lofting path does in the 2D Lofter. Lathe is used in preference to 'Rev surf' but they both perform the same action of revolving a shape around an axis.

The last Create command, *Particle Systems,* is used to create the effect of small particles, such as snow and rain, and is discussed in conjunction with animation and used in the case study Arctic Cactus.

As with 3D Studio, MAX has a variety of built-in tools that help the user to draw with accuracy and precision. Grid, 2 and 3D Snaps and Angle Snaps can all be set. MAX also allows you to create *Grid Objects,* invisible objects (when rendered) which creates a new drawing plane at a different angle than the home grid.

12.5 TRANSFORMATIONS AND MODIFICATIONS

MAX makes the distinction between Transformations and Modifications. Transformations are defined as actions which change the size, rotation or position of an object. The position of the Axis (or Transformation Point as it is called in MAX) can be important, as this is the point around which the action will take place. Icons allow the accurate placement of the Transformation Point at the default position (0,0,0), the centre of an object (or group of objects) or at any specific 3D point. If the shift key is held down while giving the move, rotate or scale commands, a copy or clone of the object is produced.

Modifications are defined by MAX as actions which change the internal structure of a piece of geometry, for example by twisting, tapering or bending it. These commands are in the Modify Command Panel. Max has a modify command, not found in 3D Studio, and that is Noise. This routine can apply random variations to the geometry to produce organic objects. Scale, Frequency and Strength can all be set, as can whether fractil algorithms are used to generate the Noise.

MAX can also create Instances and References of objects. An Instance is a completely interchangeable clone of the original mesh. If a modification is made to one of the instanced copies, all the copies change. A Reference object creates a clone dependent on the original. Modifying the original affects the clones, while modifying the referenced object has no effect on the original. Instances and references are powerful tools; change one object, and all the rest update automatically. Also, with instanced/referenced objects, instead of having say, fifty copies of the same object in a scene, you have one object with fifty 'markers' telling MAX where to repeat the object. This can save a lot of memory.

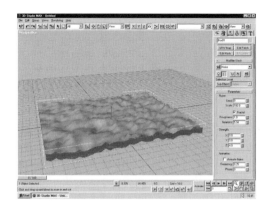

A Noise fractil algorithm applied to an object

12.6 THE MODIFIER STACK

One unique and powerful feature of MAX is the Modifier Stack. Each time a modifier is applied to an object, it is 'recorded' and put in the object's Stack. This records all the modifiers which have been applied, in the order in which they were applied. Each time the geometry is reloaded, the Stack is regenerated. The user can then go back into the Stack and change any modifier. (Any change to a modifier however, affects all modifiers above it in the Stack). Once an object is finished, and there are no more modifications to be made to it, the Stack is collapsed. Once collapsed, you can not go back into an object's Stack. Collapsing Stacks makes good modelling sense, as not having to remember the object's history frees RAM.

12.7 DIFFERENT LEVELS OF OBJECT MODIFICATION

As with 3D Studio, MAX allows you to modify whole objects or their constituent parts of vertices, faces and edges. Working at 'sub-object' level gives powerful, and infinite, opportunities for modelling. As MAX can deal with Bézier surfaces, modifications can be made to vertices or faces, and that modification can be made to affect the

The Modifier Stack dialog box

surrounding geometry. Objects can be be made more organic by 'pulling' them into shapes. These features are not contained in the Dos versions of 3D Studio and gives MAX (and VIZ) an edge in this type of modelling.

MAX divides the geometry editing commands into Edit Mesh, Edit Patch and Edit Spline to edit a variety of objects at different levels.

Within Edit Mesh, the lowest common denominator of geometry, the vertex, can be altered with Transformations, such as Move or Rotate. If Affect Region is used during a Transformation, the movement of a single vertex can be seen to alter a whole region of a mesh giving the pulling effect. The size of the area affected, and how it is affected can be set with Falloff, Pitch and Bubble.

Working at face level, a list of commands enable single or groups of faces to be: extruded, detached, deleted, moved, joined to other faces and so on. The same applies to edges with similar edit commands available as for vertices and faces.

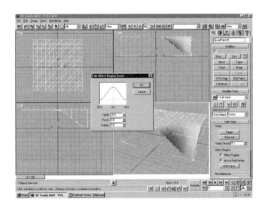

A Bézier patch

Edit Patch is designed to work with the Bézier patches that MAX can create. In reality, any object can be modified with Edit Patch simply by choosing the object and entering the Edit Patch modifier. The surface is then controlled by a structured lattice that appears over the surface. This lattice can be controlled and adjusted in a number of ways, and these alterations are reflected in underlying geometry. This is great for organic forms and is much more flexible at this type of task than anything available in 3D Studio.

The Edit Spline section of MAX contains all the tools to manipulate the 2D lines which are usually used for paths or shapes for Lofting. There are a variety of commands that enable the splines to be adjusted at both vertex and segment level, making any 2D shape possible. These commands, while given in a different way through dialog boxes, are the same as those described in the section on the 2D Shaper, in 3D Studio.

194

12.8 CREATING SCENES, CAMERAS AND LIGHTS IN MAX

Setting the scene, lighting it and adding cameras in MAX, follow many of the same conventions that are used in the 3D Editor in 3D Studio. A huge variety of lighting effects and camera views are possible.

12.8.1 CAMERAS

Two types of cameras are available, a Target camera and a Free camera. The Target Camera has a camera point and a target point and each can be defined and moved independently. They have the same adjustments of FOV, Stock Lenses, Environmental Ranges etc as those in 3D Studio. Some of the adjustments are a little different. For example, an Orbit Camera control button allows you to change the position of the camera by clicking and dragging, in any direction in the camera viewport. The Free Camera is the same as a Target Camera but without a target and is used mainly to attach to a motion path during animation.

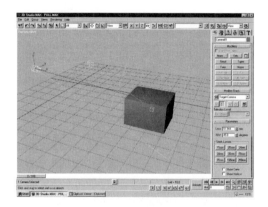

A Target Camera in the Perspective viewport

12.8.2 LIGHTING

Lighting a scene in MAX follows all the same conventions that were explored in the section on 3D Studio, though with some minor differences. The first thing to consider is MAX's Interactively Rendered Viewports. If this function is on, MAX uses two simple default light sources until other lights are created in the scene. The default lights are then turned off, and the scene illuminated with the new lights. It needs to be noted that the lighting in an Interactive Renderer, like the image itself, is only a preview and approximates

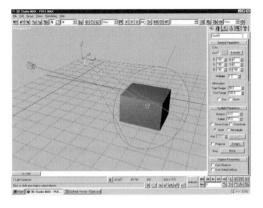

A Target Spot Light
in the Perspective viewport

The Materials Editor
interface

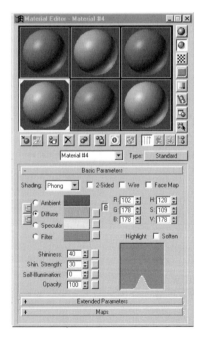

the lighting. A full render is needed to view the full affect of the lighting.

MAX provides Ambient, Omni, Directional, Target and Free Spotlights. Ambient and Omni are the same as those explained in 3D Studio. The Directional light casts parallel light rays, a infinite distance from a single point (like the Sun). Unlike the Omni light, it can cast shadows and therefore has the perimeters associated with shadow casting of hotspot, falloff, shadow maps etc. A directional light can be set to ray trace shadows.

The Target spot has a 'light point' and a 'target point' (similar to a camera). A Free spot has no target point and shines in a defined direction into a scene.

12.9 MATERIALS

As was discussed in the section on materials in 3D Studio for Dos, a 'material' is defined as a set of parameters that is assigned to the surface of a mesh to give it a quality, or to simulate a real material surface such as wood.

The Material Editor in MAX is a large window which appears within the standard MAX interface, and works in a very similar way (with a number of additions and variations) to the Materials Editor Module in 3D Studio for Dos. It has the material sample, windows (which are all rendered and not just the active one as in 3D Studio), the material libraries and the adjustments for colour and surface parameters, as seen in 3D Studio. Buttons give access to a variety of dialog boxes which set up the myriad of parameters that make up the complex surfaces that are possible. The Material/Map Browser is used to load existing materials, or create new materials based on existing templates. A browser accesses existing libraries and can display materials and maps as rendered sample spheres for easier selection, not just a list of names. There is the new feature of 'hot' materials. When the hot material is assigned to a scene, any parameter changes made in the Material Editor are automatically reflected in the scene.

Once you begin to explore the Materials Editor in MAX and begin to understand the enormously complex surfaces that can be created, the potential of the MAX Materials Editor is revealed.

MAX defines materials as Standard and Mapped. Standard materials have no surface texture or pattern and are simply a combination of colour, transparency, shininess and reflection (like paint, plastic, glass etc). Their parameters are set in the Basic Parameters dialog box, which opens by default with the Materials Editor window. Included in the basic setups is a new parameter called Filter. This defines the colour of the shadow cast by a transparent material, such as coloured glass when it is illuminated by a ray traced light. (An effect that 3D Studio cannot handle).

An Extended Parameters rollout, from the Basic dialog box, gives another layer of control over, in particular, the nature of transparent objects. How the colour of objects appear behind a transparent object, the falloff of transparency (towards the edge or towards the middle) and the Index of Refraction (IOR) can be set to affect the distortion of objects that are behind the transparent material in a scene. (Air has a IOR of 1, glass of 1.5 for example) 3D Studio does not have this facility.

Mapped Materials use the same basic principles of mapping explained in 3D Studio. A bitmap, or number of bitmaps, are applied to the surface of the material to build up pattern, texture and surface qualities. MAX offers eleven types of bitmap, (Ambient, Diffuse, Specular, Shininess, Shine strength, Self-illuminating, Opacity, Filter, Bump, Reflection and Refraction). These bitmaps can be further manipulated and defined at map level, for example noise and blur can be added to each map. An AVI or Flic file can even be applied as a bitmap and 'played' within a static or animated scene, to give the impression, for example, of a surface moving, or a TV being on.

All bitmaps used to define a material must be stored in a bitmap map path so that MAX can find them at rendering. This is usually set up at installation but can be changed and added to at any time. How the maps are applied, is controlled by the familiar mapping coordinates. MAX offers seven types of these - Lofted, Object Created, Planar, Spherical, Box, Cylindrical and Shrink Wrap, a new very useful type of mapping coordinate, which wraps a skin onto irregular organic surfaces, such as animals and plants.

Materials can be assigned to single objects, selections of objects and at sub-object level (i.e. to a face or selection of faces). They can be Blended, Double Sided (a different material to each side of a face) Checkered, Mixed and Masked (to block a section of a material with another). There are also 'procedural' materials (materials based on algorithms and not bitmaps; marble is an example in MAX), Noise (which applies random areas of two colours) and Flat Mirror for reflective surfaces. The Reflect/Refract map type, that MAX uses to simulate non-flat reflective surfaces, uses the same technique as 3D Studio. It takes six views (as if from inside the object) and applies them as a Spherical Map to the surface of the object. (See the vase example on the front cover.)

12.10 RENDERING A SCENE

The Render dialog box

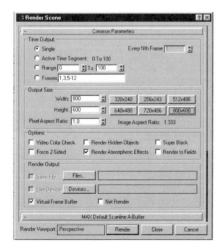

Any active viewport or region of it, can be rendered. The ability of MAX to work with Interactively Rendered Displays, often means that test renders are not as important as when using 3D Studio. (They are a very good idea, nonetheless, for complex images and lighting.) All the parameters are set in the Render Scene dialog box, similar to that found in 3D Studio. The size of the image, the use of Mapping, Shadows, Reflect/Refract, Force Wireframe, Anti-Aliasing, Pixel Size etc, can all be set. The File type and location where the rendered image or animation is to be stored, are defined and the program set to render. There is the option to save still images as GIF, Encapsulated PostScript, JPEG, Targa, and Tiff.

12.11 ENVIRONMENTAL EFFECTS

A powerful addition to MAX (and VIZ), is the increase of environmental effects such as Volumetric lights, Volumetric fog and Environment Mapping. Volumetric light appears like a spotlight shining through fog or dust. While computationally intensive and slowing rendering dramatically, they are used in scenes to create mood and to add realism. Parameters such as colour, where the fogging starts and ends, density and so on, can be set. Volumetric Fog can be set up in a similar way to its light namesake. Unlike the fog command that is is 3D Studio, the Volumetric version gives the fog irregularity and makes it look more realistic.

Environmental Mapping enables a bitmap to be mapped onto a background. For example, in an animation, to keep a sky background from moving with the camera, you can use a spherical environment map, which puts a map on the inside of a infinitely large sphere that surrounds the scene. They can also be planar. These are very good for imported images of real backgrounds, into which you wish to position, or perspective match, a newly created scene.

12.12 SPECIAL EFFECTS

Another attraction of MAX for animators and game developers is its range of Special Effects. A number are included within MAX and are much more extensive and easily used than the few provided in 3D Studio. The 'space warps' (as they are called) are Ripple, Wave, Wind, Deflector, Bomb, and Gravity.

Some of these effects are illustrated overleaf. Ripple creates concentric ripple effects, such as a ripple on the surface of water. The Amplitude, wave length, phase and decay are just some of the parameters. Wave creates a wave-like effect across any deformable object. The parameters are basically the same as the ripple parameters. The Bomb space warp is used to explode one or more objects into many smaller objects. The strength, gravity (how gravity affects the exploding particles), chaos (the randomness) and detonation (the frame number where the bomb explodes) are all parameters of bomb. Displace is an interesting and powerful space

199

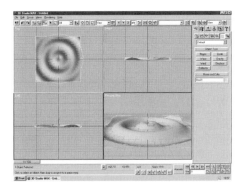

Ripple Space Warp

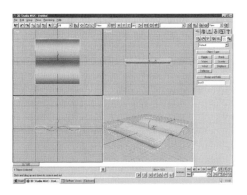

Wave Space Warp

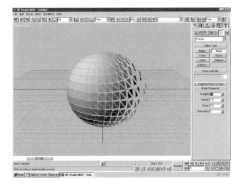

Bomb Space Warp

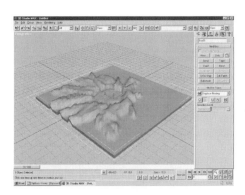

Displace Space Warp

warp, as it will modify the shape of an object using a bitmap. It is similar to a bump map, but instead of just changing the light on the surface of the material, it actually deforms the geometry based on the intensity of the light in the image.

For all the space warps to effect the geometry, there must be enough faces to show the effect. This may mean that the geometry may have to be altered (i.e. to be tessellated).

All space warps are created in a similar way. An area is defined with the mouse in a viewport and this creates a space warp icon. All the numerous parameters are set. The object is then 'bound' to the space warp with the Bind To Space Warp tool in the main tool bar. If the parameters in the object are suitable (i.e. it has enough faces to show the effect) the two objects 'bind' together. The object will then adopt that action and display it, either in its geometry (in the case of Displace) or when animated against time (in the case of 'explosion').

Morphing works in MAX in exactly the same way as in 3D Studio.

12.13 PARTICLE SYSTEMS

The particle systems Snow and Spray are also provided. They are used to create the effect of many small particles in motion, such as snow, smoke, rain, etc. These are found in the Geometry drop-down list in the Create command panel. An emitter (a rectangle with an arrow to indicate the side from which the particles appear) is created in a viewport. Particle Systems have many parameters - the number of particles, the size of the flakes, how it tumbles, when it falls, whether it is displayed as flakes, dots or ticks, what frame it begins and so on. Particle Systems can be made to interact with some of the space warps for those 'gee whiz' animation effects.

12.14 ANIMATION WITH MAX

As MAX is targetted at animators and 3D game producers, the animation controls are comprehensive and numerous. The controls are more complex and sophisticated than those available in the Keyframer, the animation module in 3D Studio for DOS. Just about any geometry, camera or modifier can be animated with many of their parameters seeming to be almost infinitely adjustable. MAX does however use exactly the same keyframing principle for animation as 3D Studio. Instead of moving into the Keyframer to create animations, you move to the desired frame and turn on the Animate Button. The button turns red and the border around the current viewport turns red to indicate that MAX is in animation mode.

Track View (called Track Information in 3D Studio) contains all the information which enable keyframes to be edited along a time line. Through this dialog box, the speed, motion and spacing of keyframes can be controlled, as well as creating, deleting, moving and copying keys. There are a large number of control commands that give instructions for these often complex animation tasks. Within Track View, different modes achieve different levels of functionality. In one mode, only keys can be edited, in another, only time can be adjusted. Each displays its own version of the Track View dialog box.

The Track View and the keyframes recorded on it, are the framework of the animation. Because of the regular way the program approximates and averages the motions and transformations between each of these key points, further control over the motion is needed to give 'natural', often irregular, motion. Precise control over this movement is achieved with Motion Controllers, (which control Tension, Continuity and Bias (TCB), Ease to, Ease from and so on) and Function Curves which chart the values of keys and the interpolation values between keys as a graph. Noise can be added to the motion to give it a randomness. Function Curves are not features in 3D Studio.

As one of MAX's most popular commercial uses is for character animation, Hierarchical linking and Inverse Kinematics can play an important role in the creation of an animation. This is an area that MAX deals with better than many other animation packages.

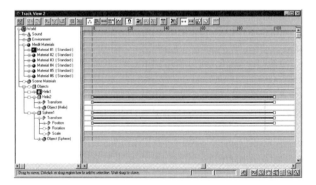

MAX's Track View dialog box

Setting up MAX to render animations, as with still frames, is through the Render Scene dialog box. The rendering parameters and file outputs can be set as for a still image, with the additional 'moving' image formats of AVI and Autodesk FLC available.

With an animation, both the amount of time and the number of frames need to be set, bearing in mind the intended output (i.e. basic smooth motion (15 FPS) film (24 FPS) PAL (25 FPS) and so on). The more sophisticated Windows NT platform makes the networking of machines together for rendering much more efficient than with Dos. As an animation is being processed in a network, the next frame to be rendered is sent to the next available machine for processing. Two similarly configured machines cut the animation rendering times in half, three cuts the time by three and so on. Animation houses often have 'farms' of machines networked together. The literature which comes with MAX claims 10,000 machines can be networked together, even over the Internet, all working on the same rendering task and only using one copy of MAX!

CHAPTER 13

3D STUDIO VIZ

Another recent addition from the Kinetix stable is 3D Studio VIZ. This is directed at Architectural, Civil and Industrial designers. It is based on the same object-orientated architecture as MAX, and is a slightly 'lighter', parallel product. VIZ will accept the same plug-ins as MAX and has file transferability with all versions of 3D Studio.

VIZ has exactly the same hardware requirements as MAX, running on Windows NT or 95. It has the same user interface as MAX, and the majority of the same functions and commands are given in the same way. However there are some differences which reflect the different target user groups.

The 3D Studio VIZ interface

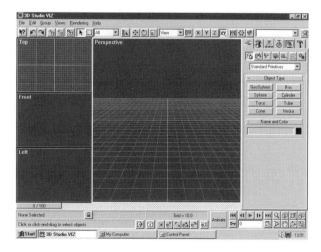

13.1 CAD

VIZ offers a greater integration with CAD systems. It freely passes information between most CAD programs using standard formats such as DXF and DWG. It also exchanges files with AutoCAD (R14) using a new compatible file format and can exchange files with

205

MAX and 3D Studio for Dos. VRML (for 3D on the Internet) and STL (for output to rapid prototyping technologies such as stereo lithography) are additional geometry file formats.

A new Section tool, which is not in MAX, can take multiple 2D cross-sections through 3D forms. This is a useful tool for Architecture and 3D modelling applications, as these sections can then be exported to CAD or technical illustration software for further refinement or output. There are also the more 'CAD-like' 2D editing commands such as trim, extend, fillet and chamfer, also not in MAX.

In VIZ there is a really extensive set of object snaps, devoted to mesh geometry. The object snaps of edge, face, vertex, endpoint, midpoint, 2D intersection, tangent and centre of face are added to the basic ones found in MAX. Snap points can be used as the centre for any transformation (i.e. rotation or scaling) and 3D objects can be positioned very accurately ensuring much more precise alignment and connection.

The Object Snap dialog box with 2 and 3D object snaps

13.2 MODELLING

VIZ shares the majority of its basic modelling functions with Dos versions of 3D Studio (i.e. primitive creation, lofting, boolean, deformation tools etc.) and has the additional functions, introduced in MAX, of Bézier patches, Edit Mesh, the Modifier Stack etc. However there are some differences. VIZ does not have the extensive range of spacewarps that are standard in MAX and therefore not the same modelling opportunities.

13.3 BACKGROUNDS

Matching a model to a existing scene can be important in Architecture. While this feature is in both MAX and 3D Studio for Dos, the Camera Match feature in VIZ is more sophisticated. Using several specified Camera Points, it can create a new camera that has the same orientation, location and field of view as the camera used to create the background image.

13.4 LIGHTS

Sunlight is a new lighting feature in VIZ. It creates realistic sunlight and shadows for exterior scenes. The sun is a directional light that follows the correct angle and movement of the Sun over the Earth. You can set the scene's geographical position in the World and the date and time. The program then calculates the position and strength of the sun, relative to the specified global position of the scene. Its movement can then be animated between a range of dates or between given hours in a day!

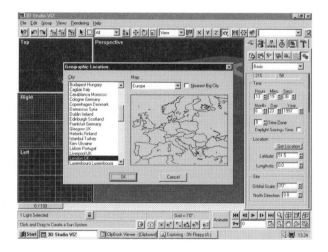

The Sunlight System interface allows you to choose any position in the world and any time of day. VIZ then creates the correct Sun angle!

13.5 RENDERING AND ANIMATION

It is in the areas of special effects and animation that there is the biggest differences between MAX and VIZ. This reflects the target user groups. All rendering setups and conventions in VIZ are more or less the same as in MAX. VIZ however, does not have such a sophisticated and powerful animator as MAX. There are no special effects as standard and it does not offer the selection of Space Warps available in MAX. Inverse Kinematics, an important feature for character animation, is also not included in VIZ.

CHAPTER 14

3D STUDIO-
CONCLUSIONS

So what are the main differences between Dos versions of 3D Studio and MAX and VIZ? There are of course the significantly different hardware requirements and the different platforms. 3D Studio needs a much lower hardware specification and is 'more straightforward' to use than MAX and VIZ. The creation of 2D lines and shapes, the passing of that geometry into the Lofter to create objects, the composition of objects into scenes in the 3D Editor and the production of an animation in the Keyframer, is all readily understandable. As there are five different modules, all using similar interfaces, each is fairly uncluttered and readable.

The MAX and VIZ interface is much more cluttered and, despite the almost exclusive use of icons, is not instantly understandable. A screen as large as possible is realistically needed to use the program with ease. On the other hand their one interface and Interactively Shaded Viewports, give a much more immediate feedback to the modelling and animation process.

Bézier patches and the ability to manipulate meshed geometry as 3D splines, is a big advantage with MAX and VIZ for organic modelling. Pulling a vertex and seeing the surrounding vertices being affected, is a powerful tool, frustratingly missing from 3D Studio. The Modifier Stack which gives the ability to go 'back into' an object's creation history, and edit each modification, is a very useful innovation. The multiple undo and redo commands, which are not in 3D Studio for Dos, make mistakes in MAX and VIZ less problematic. The surface and mapping variations are greater in MAX and VIZ.

MAX is orientated much more to animation than its partners. Its Special Effects, Space Warps, complex IK and flexible organic modeller, make it an obvious choice for the serious animator.

VIZ integrates well with CAD programs and conventions. Its efficient object snaps give great speed and control when transforming and manipulating 3D objects with precision. These features are often useful for Designers and Architects.

In the end, the decision about which 'version' of 3D Studio used, is likely to be decided by the nature of the work to be carried out, what version is affordable or availability, and access to hardware.

CHAPTER 15

CASE STUDIES

To demonstrate some of 3D Studio's modelling techniques, four images have been used as case studies. These provide the opportunity to look at a number of modelling processes used during specific modelling sessions, looking at the problems and solutions that present themselves. Within these examples many of the main modelling and rendering functions are used. The route used to achieve the results is explained, but as with most modelling projects, there are often many other ways to achieve similar results.

The 'Hall Table' image

15.1 THE HALL TABLE

The 'hall table' scene was modelled and rendered in 3D Studio Release 4. If there are a number of complex objects in a scene (like those on the table) it is usually best to create each object in a separate project and merge them individually into the scene. This technique is used in this example.

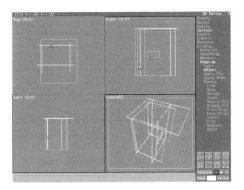

The environment

*Rendered view of
the window*

*The table in wire
frame in the 3D
Editor*

1. Firstly an environment is created in the 3D Editor. Create/box is used to make a flat form for the back wall of the scene. A white bumpy semi-matt material is made in the Materials Editor using 4% of the Sand.cel as a bump map. This material is then applied to the 'wall'. As this material uses a bitmap, the wall has to have planar mapping coordinates applied to it. By using mapping/adjust/bitmap fit (and then choosing the Sand.cel bitmap) this ensures that the mapped texture is properly 'tiled' without visible joints. This wall is then copied using object/move and object/rotate (with the shift key held down to produce clones) to form the four walls of the 'room'.

To cut a hole in the back wall to simulate the window, a box is created and positioned where the window is to be. Using create/object/ Boolean and then the subtraction option, the hole is 'punched' out. Before using the box to Boolean out the hole however, a copy is made as the original will be destroyed during the Boolean operation. On the inward face of the box copy, an image of a sky is mapped to simulate the outside world.

From the World Creating Toolkit, which ships with 3D Studio Release 4, the mesh of a set of blinds is imported, scaled and placed in front of the window. This demonstrates the usefulness of being able to import readymade meshes. To model even this simple object could take considerable time and it is only being used as a reflection in a mirror. With the environment set up, and saved as a project, it is time to create the objects in it.

2. The first object to be created is the table. Still in the 3D Editor, rectangular boxes (one 1000x60x60 for the legs and one 1000x75x20 for

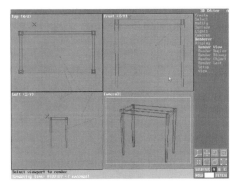

the frame) are created. The material White Ash is assigned to each object. The leg has a cylindrical type mapping applied to it, ensuring that the grain of the wood is the correct scale and in a realistic direction with test renders. Planar mapping coordinates are adjusted and applied to the table structure. Using move and clone (with snap on to keep everything aligned) three copies of the original leg are made and moved into position in the top viewport. The structure piece is also cloned, and the two ends scaled, to create the framework. The top is a simple rectangular box created in the 3D Editor and given the material Green Marble.

The finished table with and without the top

3. The lamp is next to be modelled. In the 2D Shaper, the cross section is created by first making a 16 circular sided n-gon. Every other vertex in the circle is picked using select/vertex/single and pulled in towards the middle of the circle using scale/vertex/ selected. This shape is then assigned and imported into the Lofter where it is lofted along a straight path using a Scale deformation grid to create the curve of the stem. A hemisphere is made and squashed using scale (first restricting the scaling to the Z axis and then the Y or X axis using the Tab Key) for the base.

The shade is made by first creating two concentric circles close together in the 2D Shaper. These are imported onto a straight path in the Lofter. Scaled down copies of these circles are then put at the other end of the path and the object created. As there are two shapes at each end of the path, a thin hollow tapering form is the result. The bulb and the fittings are created using a variety of lofting and lathing techniques.

The individual meshes then have materials applied. The stem and base are made a standard matt yellow. The shade is made a semi-

*The 2D crossections
for the lamp*

The finished lamp

transparent material simulating semi-opaque glass. The bulb in the lamp appears to glow. This is achieved by giving the bulb's material an amount of self-illumination.

The whole lamp model is then saved as a mesh, ready to be imported into the scene later.

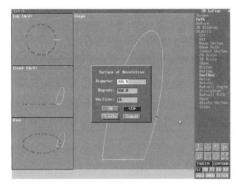

The 2D shape for the Bowl imported onto a circular path in the 2D Lofter

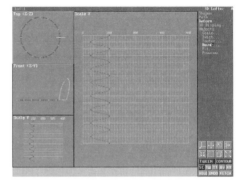

The Scale Deformation grid for the bowl object

The final object

4. The modelling of the bowl demonstrates the creative use of a Scale deformation. A simple cross section is created in the 2D Shaper (a squashed and skewed 12 sided n-gon with the end cut off it) and imported onto a circular path. The middle of the 'flat' end is aligned exactly with the path. Then using a complex scale deformation in both the X and Y axis, the distinctive rim shape is created. The base is a thin disc created and positioned in the 3D Editor and the oranges are simply two spheres. The bowl is given the material white plastic and the spheres a slightly bumpy orange material.

214

5. The mirror frame on the back wall is created from simple rectangles in the 3D Editor, which are tapered to a 45° angle at their ends. As the mirror frame has a wood material mapped onto it, each side must be made separately so that the orientation of the mapping can be different in each one. The reflective surface, which has an automatic reflection map, is applied to the front face of a thin box to simulate a mirror.

Picture frame with imported image applied as bitmap

6. The picture frame on the table is created from a cross section of the frame being lofted along a rectangular path. It is given a metal material. The image, which could be any image in any compatible file format, is scaled and mapped onto the front of a thin rectangle.

7. The vase form in the middle of the scene, demonstrates how 3D Studio deals with reflection on a non-flat surface. Using the cubic reflection map method explained earlier in the book, the program renders six views as if from inside the object. These images are then bitmapped onto the surface of the object simulating reflection. The tulips in the vase are imported as an existing mesh.

8. The cactus is also imported as a mesh from the 3D Studio World Creating Toolkit.

9. The original project, where the environment was built, is opened again and all the meshes of the created and externally source objects are imported and 'merged' into the scene using File/merge. Some time is spent arranging the objects in the scene. While all the newly created objects were built full scale and relate to each other when imported (which is good modelling practice) the externally source meshes, such as the cactus and the blind, need to be scaled to match the other objects.

Vase with cubic reflection map

10. One main shadow-casting spotlight illuminates the scene from the left. An omni light gives some fill in from the right. A spotlight is placed just above the lamp and directed up the wall. This sim-ulates light being produced by the lamp. Many low resolution test renders are needed to fine tune the lighting in the scene. A final high resolution render of a camera view produces the completed scene.

The imported cactus mesh

215

*The final rendered
image on the back
cover*

15.2 THE TEAPOT

The model of the teapot, cup and
saucer was modelled and rendered
using 3D Studio Release 4.

1.The modelling for this object begins
in the 2D Shaper. An n-gon with 10 flat
sides and a radius of 70mm is created.
This is the cross section of the main
body of the teapot at its widest point.
Pressing F2 moves the modelling into
the 3D Lofter. By using path/move
vertex, the end vertex of the path is
moved up to create a path 140 units
long. As there is going to be some
detail and change of cross section as
the form is lofted, the steps setting
needs to be turned up to 10. The 2D outline is imported using
shapes/get/shaper. The Scale deformation tool is then used to
define the outline of the form by pulling the end vertex and
adjusting it by holding down the mouse. An object/preview
confirms that the shape is correct and Object/create generates the
model and puts it into the 3D Editor.

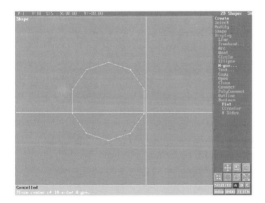

The outline for the teapot body

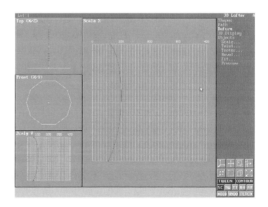

The Scale deformaton grid for the teapot body

216

2. Back in the 2D Shaper, two concentric circles are created by first drawing a circle with a 40 unit radius, and then using the create/outline tool to off-set that circle 2.5 to each side. These circles are assigned and imported into the Lofter and lofted along a short straight path with no steps. This makes a hollow ring which will be used as the lid seat.

3. Modelling the handle (and the spout) is a little more complex. As with many objects like this, it took several attempts to get it right. First the profile of the outside of the handle is created in the 2D Shaper. I choose to do this by first creating a 16-sided n-gon, scaling and then skewing it in the X,Y direction to create a slightly twisted ellipse. Some of the segments are then deleted leaving an open polygon. I find this a good method for making these types of smooth curves. (This profile could of course be generated in several different ways. For example a line with each vertex being individually splined.) Still in the 2D Shaper, the cross section for the handle is generated simply by creating a 6-sided circle of 10 units and scaling it in the X axis by 80%. The two vertices at one side are deleted using modify/vertex/delete to create the desired shape.

Firstly the handle profile is assigned (shape/assign) and imported into the Lofter using path/get/shaper. The default setting of 5 steps is fine as the number of vertices, combined with the steps, will give the necessary detail along the path. Moving into the 2D Editor, the cross section shape is assigned. Back in the 3D Lofter, that polygon is then imported as the Shape using shape/get/shaper. The cross section needs to be aligned to the end of the path so that the flatter edge is exactly on the profile path. This is best done with shape/centre and than move the shape to the left to align it exactly with the profile path.

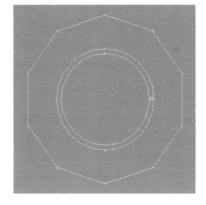

The outline for the lid seat

The outline and crossection for the handle

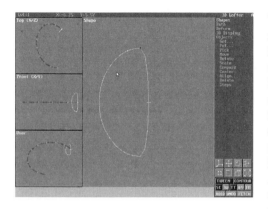

The shape for the handle imported onto the path in the 3D Lofter

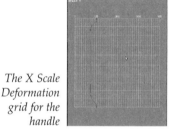

The X Scale Deformation grid for the handle

A loft preview for the spout

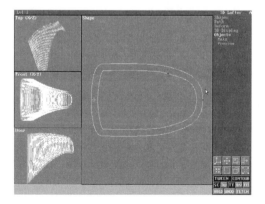

The most interesting part of this piece of modelling is how the Scale deformation tool is used. We want the handle to remain a constant width as viewed from the back but have a changing profile from the side. To achieve this the Scale deformation symmetry is turned off and left at the default in the Y scale. In the X scale, by inserting, moving and splining vertices at both ends of grid, the inner profile of the handle can be described. As the 'back' of the shape is aligned to the outside profile, the deformation will only affect the inside profile and the outside will follow the original created path. A small amount of Teeter is also applied in the X axis to refine the side profile. Several previews were necessary to fine tune the deformations and get the handle looking right.

4. The spout is modelled in a similar way to the handle. In the 2D Shaper the cross section of the back of the spout is created using a circle and square as a basis of the form then moving, deleting and adjusting vertices to create the desired shape. An outline (2mm on each side of the line) is created using create/outline.

The profile of the spout is drawn as a simple curved line using the create/line command. It is important that this line has several vertices so when it is imported as a path it will have enough interim steps to give all the necessary detail in the lofted shape. This shape is assigned and imported into the lofter as a path. The outlined shapes are then imported as 'shapes' using get/shape/shaper. The cross section is positioned so its top edge is aligned with the curved path.

218

Once again, the deformation tools of Scale and Teeter play a
significant role in the creating of this part of the model, to give the
spout its subtle shape. The Scale deformation symmetry is set to 'off'
and in the graph for the X axis, a deformation line is created and
adjusted to define the profile of the lower edge of the spout. In the Y
axis a straight scale is used and Teeter is set to give the spout a
changing lilt as it moves up the path. Again several adjustments
needed to be made to all the deformation settings before the desired
shape was achieved and exported to the 3D Editor.

5. To create the teapot lid, a cross section is drawn in the 2D Shaper
using a rectangle and an arc joined together. In the 3D Lofter this
shape is rotated using a 180° circular path (of the same diameter as
the lid) with the edge of the lid cross section aligned to the path.

6. Once all the lofted components are created (body, lid seat, lid,
spout and handle) they are manoeuvred into position and aligned
using object/move and object/rotate to construct the finished teapot.
A small smooth sphere is created in the Editor and 2D scaled to
make the knob and the whole object saved as a mesh (.3ds file).

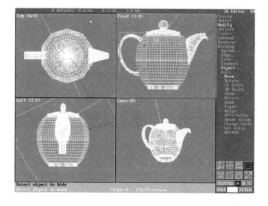

Assembling the Teapot
parts in the 3D Editor

15.2.1 THE CUP AND SAUCER

7. The main cup body is made up of two parts joined together. The faceted outside form is created by lofting and scaling a 10-sided n-gon along a straight path in a similar way to the teapot body. The round inside is modelled with a cross section (drawn in the 2D Shaper) revolved around a circular rev-surf path. This cross section needs considerable splining and adjustment of vertices in the 2D Shaper to refine the shape.

The creating and adjusting the 2D crossection of the cup in the 2D Shaper

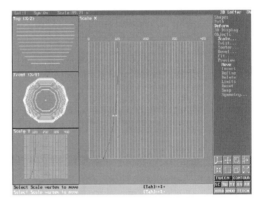

A preview of the loft for the main body of the cup in the 3D Lofter

8. The cup handle is a copy of the teapot handle which is cloned (using objects/move and holding the shift key) and scaled (using object/scale) in the 3D Editor.

15.3 THE ARCTIC CACTUS

This scene was modelled and rendered in 3D
Studio MAX. It demonstrates some of the special
effects and modelling techniques which are
available in MAX and not in 3D Studio for Dos.

1. The environment is set up and the undulating
landscape is the first object to be modelled.
Working in the Top view of MAX's default four
viewport configuration, a box is created by
dragging a rectangle to describe the top of the
box, and clicking in the viewport to define its
depth. (The command for box is found in the
Create/standard primitives command panel.) As
this object is to have a Space Warp applied to it, it needs to have
enough faces to show the effect, so in the box command rollout area
the length and width segment settings is set to 50.

*The Arctic Cactus
image*

2. From the Create command panel, the Space Warp Displace is
chosen. In the top viewport the rectangular Space Warp icon is
drawn approximately the same size as the box. In the Displace
rollout the image Cloud.jpg is chosen and the displacement strength
set to 10. From the main tool bar, Bind To Space Warp is chosen. The
box object is then picked and dragged over the displace icon until
the cursor icon changes, when the mouse is released, the
displacement map appears in the geometry.

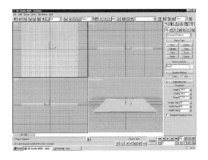

*The Space Warp Displace used on a thin
box for the surface. Note the number of
segments used to create the box*

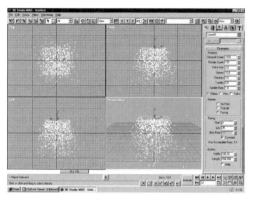

The Snow
Particle System

3. From the two default particle systems Snow and Spray that MAX offers, Snow is used to give more drama to the scene. Snow is found in the Create command panel. Of the numerous settings, only Render count (set to 10000) and Variation (set to 4) are changed. In the top viewport a rectangle is defined (which is the particle systems source and displayed as an icon) over the box. The particle system icon needs then to be moved, in the front viewport, to some distance above the box. As this effect is referenced against time, the snow is only visible when the Animation time slider is moved.

4. The scene also has some Layered Fog applied to it. This fog effect is chosen from the Add Atmospheric Effect dialog box. A rollout gives several adjustment options. In the Layered Fog controls, the bottom is set to 0, the Top to 40, the density to 60 and the Falloff to top, is turned on. This creates a layer of fog when rendered, to give the impression of fog hovering over the ground.

5. From the Environment dialog box, the image Sky.jpg is chosen and assigned as the background. A shadow-casting spotlight illuminates the scene from the left and a couple of omni lights are added. This completes the environment setup.

Atmospheric Effect
dialog box

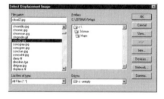

Environment dialog box

The Atmospheric Effect Fog, the
Particle System Snow and the
Environment Map Sky.jpg
added to the object 'box' to
create the environment

6. Next to be modelled was the cactus. A 2D star is drawn (found in the shapes command panel) to act as a basis for the cross section of the cactus. Each of the vertices are then selected and splined to create a smooth undulating outline. A straight line is drawn to act as a path for the loft.

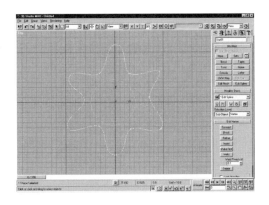

The 2D crossection

7. From the Create command panel, the Loft Object option is chosen. The straight line is assigned as the Path, the star shape as the Shape and the object is automatically lofted. To give the object a more realistic irregular form, a Deformation Curve is applied to the lofted object. From the Deformations Rollout in the Modify command panel, Scale is chosen. This displays a dialog box. Several Bézier points are added to the Deformation Curve. They are then moved and splined to create an irregular profile.

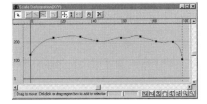

The scale deformation for the cactus

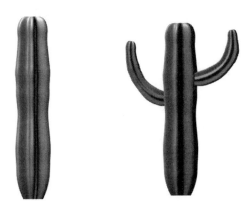

8. The main body of the cactus is then copied twice. These copies are Scaled, Bent and positioned to form the two arms of the cactus. The whole object is then assigned a green matt material.

9. The final object created is the igloo. This is made from a simple extrusion which forms the door, and a hemisphere for the main dome. The dome has a white brick material (made from the standard material Bumpy Brick with the colour changed to white) applied as a Spherical mapping. The doorway has a Planar mapping of the same material.

10. All the objects are positioned in the scene and the scene rendered.

The finished igloo with the White
Bumpy Brick material mapped onto it

15.4 The 3D

The '3D' image demonstrates more of 3D Studio's material and modelling capabilities. The 3D text is created in the text editor and extruded. Some scaling and modelling gave the '3' a constructed feel and a 'mouse' is modelled. A variety of materials (wood, metal, glass and plastic) are assigned to the different parts of the objects and they are all placed on a reflective surface. The image is then rendered twice from the same camera view and at the same size. One rendering is in full colour using Metal shading, the other has the 'force 2 sided' option turned on in the Rendering dialog box to render the scene as a wire frame. The two rendered images are then exported to Adobe Photoshop. The wire frame image is put on a layer below the shaded image and part of the shaded image is 'erased' to expose the wire frame. A soft edge gives the impression of the D fading from solid to wireframe.

Wire frame image

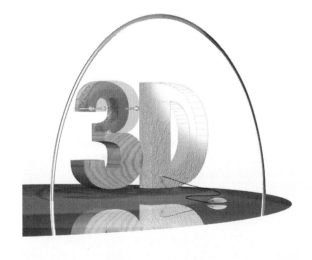

Rendered image

Merged image

225

APPENDIX 1

GLOSSARY OF TERMS

1-Dimensional When referring to spatial dimensions: having length but no breadth, such as a straight line.

2-Dimensional (2-D) When referring to spatial dimensions: having two dimensions (length and breadth, or length and height), such as a plane.

21/2-Dimensional (21/2-D) Usually referring to an animation created in several flat layers to give some of the depth effects of true 3-D.

3-Dimensional (3-D) When referring to spatial dimensions: having three dimensions (length, breadth and height), such as an object.

3-Space Three-dimensional space.

4-Dimensional Usually referring to three spatial dimensions plus the added dimension of time.

AI Artificial Intelligence is involved with building features associated with natural intelligence into machines.

Aliasing (See Spatial aliasing, Temporal aliasing)

Anti-aliasing The removal of aliasing artefacts, most commonly involving the smoothing of jagged edges on output displays.

Artefact Used to describe some part of the image which has been inadvertently created, or is unsatisfactory as a result of deficiencies in the system, and constitutes an error.

ASCII (pronounced 'askey') An internationally agreed set of characters as produced by a standard keyboard.

Axis   The line about which an object rotates.

B-rep   Boundary Representation method for creating objects by defining a polygonised surface mesh.

Back-face culling   A simple, but crude, hidden line removal method.

Bézier   Invented a mathematical description of a curve, based on the definition of a few points or use in the car industry. It is widely used in computer graphics, often in a context where the curve is to be 'tuned' interactively, to create 2-D lines, 2-D and 3-D paths and 3-D surfaces.

Bicubic patch   A means of describing a curved surface using cubic functions. A surface may be divided into a number of patches with suitable continuity at boundaries.

Bit   From 'binary digit'. The basic unit of computer information (which can be represented by either 0 or 1).

Bit map   The representation of the screen image in memory, stored as pixel intensities.

Boolean operations   Operations based on the logical relationships of AND, OR and NOT (union, difference and intersection). In CSG modelling, for instance, the logical operators can be used to join or cut existing objects into new objects.

Bounding box   A simple space frame which can act as a tempo rary substitute for a more complex object in order to simplify calculations when a quick approximation (of a movement, for example) is needed.

Buffer   An area of memory (which may be internal or external) temporarily reserved to hold information which is currently required. A frame buffer, for example, holds the displayed image as a matrix of intensity values.

Bump mapping   By perturbing surface normals across a flat surface, suitable rendering algorithms will produce what appears a bumpy surface.

Byte   A set of (usually eight) contiguous bits.

Cartesian coordinates   Two-dimensional points which can be located by reference to calibrated horizontal (X) and vertical (Y) axes. In three dimensions an additional axis (Z) establishes depth.

Canonical position   The expected, default position of an object on creation and before it is moved. Normally centred on the origin and with key facets orthogonal to axes.

Character Animation   Animation that includes the assignment of anthropomorphic motion to characters or inanimate objects.

Clock rate The rate at which operations are carried out by the CPU.

Constraints Limitations applied, particularly to the movement of an object.

Continuity The degree of smoothness with which line and surface sections join.

Convex hull The 'skin' created by enclosing all the extreme points of an object.

Coordinates (See Cartesian coordinates, Polar coordinates and Spherical coordinates.)

Cosine shading (See Lambert shading.)

CPU Central Processing Unit. The heart of the computer.

CSG Constructive Solid Geometry. A modelling method in which primitives, such as cubes and spheres, are combined using Boolean operations.

Data compression The algorithmic reduction in size of data files.

Default state The state in which something will exist until it is consciously changed.

Degrees of freedom of movement The number of singular ways in which an object can move. For example, a particle has three degrees of freedom of movement (along X,Y or Z), a rigid body has six (along X,Y and Z plus rotation around X,Y andZ).

Desktop A visual metaphor used in a WIMP environment whereby the VDU screen is organised as if it were a real desktop.

Digitiser (3-D) A device for acquiring and inputting spatial data about the surface of an object.

Digitising tablet An input device with a flat, sensitive surface which can be drawn on with a stylus in the manner of a pencil and paper. A puck may be used to acquire 2-D coordinates from drawings aligned on the tablet.

Disc drive A secondary storage device in which data are saved on a removable rotating disc (which can be conveniently stored or used to transfer data between computers). The most common storage medium is still magnetic.

Dithering One means of simulating a larger palette of colours than is actually available.

DMA Direct Memory Access.

DOS Microsoft Disc Operating System

DRAM Dynamic Random Access Memory.

DTV Desktop Video.

DVI Digital Video Interactive technology.

Dynamics The branch of mechanics dealing with the way masses move under the influence of forces and torques. In creasingly used to drive animations by the application of physical laws.

Element A group of faces in 3D Studio that share common vertices and may make up a larger object.

Expert system Provides a means of solving 'significant' problems by applying rules to a data bank of relevant information culled from human experts in the field concerned.

Extrusion A swept surface method of modelling, where a 2-D template is dragged through 3-space along a path, in the simplest case at right angles to the plane of the template.

Face (or facet)A planar surface which constitutes one face of a polygonised model.

FFD Free Form Deformation. A method of deforming an object by applying transformations to a cage of control points. Objects can be created by using FFDs on primitives, and the same principles can be used to animate a change of shape.

Firmware The embodiment in hardware of a function normally associated with software. For example, some frequently used rendering algorithms might be built into a chip in order to gain substantial speed increases.

FLOPS FLoating Point Operations per Second.

Fractal A term used to describe the self-similarity of some phenomena when viewed at different levels of detail. The principle is typically used in computer graphics to generate mountains, clouds and such like, from a very small data base.

Frame buffer A piece of specialised memory (which may be internal or external) reserved to hold one or more images for quick access and/or processing.

Grid A regularly spaced set of markers that can be displayed for reference while drawing in a computer graphics program.

GUI Graphical User Interface.

Gouraud shading A shading model which improves on Lambert shading by smoothing intensity across surfaces.

Granularity A rough description of the level of detail at which an operation is conducted (e.g. rough-grained = low level of detail, fine-grained = high level of detail).

Hard disc A sealed unit for secondary storage which is costructed internally like tiers of disc drives. It combines larger storage space (typically) and relatively quick access.

Hardcopy Output in permanent form such as on paper or film.

Hardware Refers to the physical components of a computer system i.e. the boxes that sit on your desk.

HCI Human Computer Interface. The boundary between the machine and the user at which they communicate with one another.

Hidden line/surface The removal of lines or surfaces which we would expect to be obscured when viewed from a specified direction.

High level In this context, a high level operation is one in which the operator does not need to involve himself with the details of how the operation is carried out. (See Low level.)

IFS Iterated Function System. Used to derive a simple set of fractal rules from complex data, such as an image, and thus allow potentially extreme data compression.

Image mapping A means of applying a picture to a surface or of wrapping a picture around an object.

Interactive Allowing the user to respond to the running of an application with fresh input while it is in progress.

Interlaced A raster scan in which alternate scan lines are re freshed on each pass. This means that less information needs to be handled at any one moment than in a non-interlaced scan but that the complete image is refreshed less often.

Inverse Kinematics (IK) Used extensively in character animation to control linked objects in a hierarchical chain where if you transform the child object the parent object is affected. It is also used to set real world physical constraints on the movement of animated joints to give greater realism to character motion.

I/O Input/output.

Iteration Repetition, usually of a piece of a computer program (in which case it is called a loop). Something that computers are particularly good at doing.

Jaggies An informal term for the jagged lines in a pixelated image which it is usually desirable to minimise.

JPEG algorithm Joint Photographic Experts Group algorithm for data compression of still images.

Keyframing   A term that comes form traditional cel animation. The master animators would draw the 'key' frames and the assistant animators the 'tweens'. In computer animation, the operator creates the keyframes and the tweens are handled by the program.

Lambert shading   A basic shading model in which each facet is evenly shaded according to the angle at which the light hits it.

Lathe A term sometimes used instead of 'spin'.

Lofting   The term used in 3D Studio for the sweeping of a shape along a path to create a object or connecting cross-sections through an object by triangulating a surface between their edges.

Low level A level of operation where the operator is required to become involved with the detail of the machine or process. (See High level.)

Mach banding   A phenomenon in which a smoothly shaded surface appears to have dark streaks on it.

Mapping   (See Bump mapping, Image mapping, Reflection mapping,Texture mapping.)

Mesh (wireframe)   Most common term used in 3D Studio for a wireframe object.

Metal Shading   Used by 3D Studio, it is similar to Phong shading but produces a more metallic effect by using different types of specular highlights.

Metamorphosis A change of physical form, often easily animated by interpolation between the start and end forms.

MIPS   Million Instructions Per Second.

Modelling The construction of objects in a scene prior to rendering (or choreographing movement).

Mouse   A common input device which fits in the palm of the hand and is rolled over a flat horizontal surface to control the movement of a screen cursor.

MPEG algorithm   Motion Picture Experts Group algorithm for data compression of motion picture images.

Multimedia   A term used to describe the mixed use of still and moving visual media, together with sound, normally under the control of a computer.

Multi-tasking   The ability of some computers to work on several tasks at the same time. In fact, although things appear to be happening at the same time, the machine is normally avail able CPU time. Not to be confused with parallel computing.

N-gon   **A 2D polygon with any number of sides of equal length.**

Non-interlaced   A raster scan in which each scan line is refreshed on each pass (See Interlaced).

Noise   Applies random variation to geometry to produce organic-looking objects or movement.Fractal algorithms can be used to generate noise.

NURB   Non-Uniform Rational B-spline. A type of B-spline which is particularly flexible in interactive use.

Normal   (See Surface normal)

NTSC   Broadcast standard used in USA and Japan.

Object orientated   Describes a type of programming language, growing in popularity, in which program elements are considered as separate objects which can communicate with one another. The term is also used to refer to an image which is defined as a number of separate parts and their relationships to one another (as opposed to a raster image).

Object space   The 3-D space of the object's world.

Origin   The point at the centre of a coordinate system where X,Y and Z all equal zero.

OS   Operating System.

Painters' algorithm   A simple method of removing hidden surfaces by overpainting.

Paint system   An image creation system which simulates traditional materials used for drawing and painting

PAL   Broadcast standard used in much of Europe.

Palette   The range of colours available. Dependent on hardware and software constraints, the range can stretch from 2 to more than 16,000,000.

Path   The course along which something moves.

Parallel architecture   A design of computer in which a number of tasks can be carried out simultaneously, i.e. in parallel'. Some times referred to as 'non-von' since it is a departure from the traditional von Neumann computer architecture.

Parallel processing   The simultaneous processing carried out in a parallel computer (See Parallel architecture).

Particle   A single point in 3-space. Theoretically infinitely small, but often treated as a small mass limited to three degrees of freedom of movement.

233

Particle system A system containing a number of particles (typically between ten thousand and a million) which might be used to model 'soft' objects or to animate flow through a medium, for example.

PC Personal Computer. Tends to be used to refer to IBM compatible machines.

PDL Page Description Language, e.g. PostScript.

PHIGS Programmers' Hierarchical Interactive Graphics System. A 3-D graphics standard established by the American National Standards Institute (ANSI).

Physically based modelling The representation of a model in terms of its physical attributes, such as mass, and forces. Such a model can be controlled by the application of the laws of physics and is thus ideal for simulation.

Phong shading A smooth shading method which incorporates specular highlights.

Pixel From 'picture element', the smallest element out of which a screen display is made.

Plotter An output device in which a pen, or selection of pens, is raised and lowered whilst being carried across the surface of a piece of paper. Traditionally associated with engineering and architectural drawing.

Polar co-ordinates A point in two dimensions can be defined by its distance from the origin and the angle between the positive X axis and a line from the origin to the point.

Polygon A planar figure bounded by straight sides. (Also used in 3D Studio to describe any 2D line or shape).

Precision errors Errors arising from the inability of the computer accurately to store numbers beyond a certain length. If the result of a calculation is a long decimal number the machine might need to truncat it for storage, thus introducing a small error which could become exaggerated in further calculations.

Primitive A simple object (such as a cube or sphere) which is provided as a basic 3-D unit in a modelling system.

Procedural Materials (Also known as solid materials) Materials generated by mathematical algorithm. With this type of material, you can cut out a part of a mesh object and the material will still look correct (unlike a mapped material). Marbles, woods and granites are popular procedural materials.

Puck A device similar to a mouse but with a cross-hair sight for accurate alignment, used for the input of points (for instance from a drawing). (See Digitising tablet)

Radiosity An effective but ponderous shading method which is particularly good at dealing with diffuse light.

RAM Random Access Memory.

Raster image Often used to describe a pixel based image (in which the image is recorded as a collection of pixel intensities) as opposed to one which is vector based (and can therefore be displayed at the best resolution of the output device).

Raster scan The scanning of a monitor screen by an electron beam.

Ray tracing A simple, though time-consuming, rendering method which produces 'realistic' shadows and reflections.

Real time A one to one relationship between display time and real-life time.

Reflection mapping A means of applying a picture of an object's surroundings (or imaginary surroundings) to its surface in order to simulate reflection.

Refresh rate The rate at which an image is redrawn on a screen.

Render To make the internal mathematical model of a scene visible. Usually refers to the algorithmic realisation of the effects of lighting, surface colour, texture, and reflection.

Resolution Although a number of factors affect resolution, it is generally taken to describe the apparent level of detail an output device is capable of resolving.

RGB A colour system where all colours are defined as a mixture of red, green and blue (as in a TV).

RISC Reduced Instruction-Set Computing.

ROM Read-Only Memory.

Scanner A 2-D image input device which scans an image (using similar technology to a photocopier).

Screen space The two dimensional space of the screen image. (See Object space)

SECAM Broadcast standard used in France, Russia and elsewhere.

SIGGRAPH ACM (Association of Computing Machinery) Special Interest Group in Graphics.

SIMD   Single Instruction Multiple Data. An architecture for parallel processing.

Snap Makes the user's mouse actions 'snap' to fixed, evenly spaced positions to ensure accuracy while drawing or moving lines, shapes, or objects.

Soft modelling   The modelling of non-geometric, often natural, forms.

Software   Refers to the programs, expressed in machine readable language, that control the hardware.

Solid texture   Texture pattern running right through the volume of the object rather than just on its surface.

Space Warps   A term used in 3D Studio MAX for the special effects (such as ripple, displace and wave) which affect the 3D space in and around an object to affect how it appears or behaves.

Spatial aliasing   A problem of discontinuity arising from trying to match correct locations to the nearest available point on an output device. See jaggies.

Spatial occupancy enumeration   A volume modelling method in which an object is defined by the presence or absence ofvoxels.

Spherical coordinates   An extension of the polar coordinate system which deals with 3-D by incorporating an extra angular measurement.

Spline   A flexible strip of wood used to create smooth curves (originally in shipbuilding), the same result is now achieved mathematically.

Spinning (Rev-Surfing)   Process of creating a swept surface by rotation of a 2-D template around an axis.

Staircasing (See Jaggies)

Stochastic   Random within prescribed limits. Stochastics are often employed to produce variations on a basic theme.

Stylus   A pen-like device used in conjunction with a digitising tablet, mainly used for the freehand creation imag creation.

Sub-pixel Theoretical division of a pixel into smaller units for the purpose of calculations.

Super sampling   Conducting calculations at a finer resolution than the output device will be able to implement. Used as a means of dealing with aliasing by sampling at sub-pixel level.

Surface normal   A vector orthogonal to a surface. Central to many computer graphics calculations.

Swept surface   A 3D surface created by passing a 2D template through 3D space

Temporal aliasing   A problem of discontinuity arising from trying to match accurate moments in time to the nearest available time-point on an output device.

Teleological modelling   An extension of physically based modelling to include goal-orientation. The attributes of an object include a knowledge of how it should act.

Texture mapping   Used to describe both the wrapping of a 2-D representation of texture onto a surface in object space (although this might be better referred to as image mapping) and the transfer of an external bump map to a surface.

Texture space   The space inhabited by the 3-D textural infor mation used in solid texturing (where the texture runs through the object like grain through wood).

Texel   TEXture ELement.   A single unit of texture (which might be compared with a pixel or a voxel).

Transformation   The alteration of shapes or objects by applying geometrical rules to their coordinates, e.g. translation (movement in a straight line), scaling and rotation.

Transputer   A chip for parallel processing, containing its own memory and processing unit.

Triangulation   Division of a surface into triangular facets. The division is often required because a triangular facet is necessarily planar, and non-planar facets would create problems during other calculations.

VDU   Visual Display Unit. Normally refers to the monitor screen.

Vector   Usually refers to the storage of image data in terms of relative measurements (which can therefore be displayed at the best resolution of the output device) as opposed to storage of an image in terms of pixel intensities. In mathemat ics a vector is a value having magnitude and direction, and in modelling a point can be represented by a vector and trans formed using matrices.

Vertex   A point in 2-D or 3-D space which is connected to others in order to build shapes or facets.

Viewing transformation   The mathematical conversion of 3-D information so that it can be presented in 2-D, as if viewed from a given point (with perspective).

Virtual   Appearing to be something it is not. Hence virtual memory describes the use of secondary storage as if it were main memory and virtual reality describes a simulated situation which aims to be indistinguishable from one of real life.

Visualisation   Making complex information (often being large quantities of scientific data) understandable through presentation in a visual form.

Volume visualisation   The rendering of 3-D volumes by voxel methods (See Spatial occupancy enumeration).

Von Neumann architecture   The traditional computer architecture in which operations are carried out sequentially (as opposed to concurrently).

Voxel   From 'volume element'. A cubic unit of 3-D volume defined at a size appropriate to the required resolution, sometimes described as the 3-D equivalent of a pixel.

VR   Virtual Reality (See Virtual.)

WIMP   Windows, Icons, Menus and Pointers used in an interface.

Wireframe   A representation of an object using only the edges of its constituent polygons.

WORM   Write Once Read Many. Refers to a storage device from which information can be read but to which it cannot be written.

WYSIWYG   What You See Is What You Get. Describes a system where the screen representation exactly represents the hard-copy output. The two are otherwise often not the same since the device resolution determines how accurately images can be shown and this represents a common problem in many graphics situations (such as DTP).

X-axis   The horizontal axis in a Cartesian coordinate system.

Y-axis   The vertical axis in a Cartesian coordinate system.

Z-axis   The axis representing the dimension of depth in a 3-D Cartesian coordinate system. In its most usual presentation the Z-axis can be thought of as going back at right angles to the vertical plane on which the X- and Y-axes exist. (This is a left-handed system. In a right-handed system the Z-axis would come forward from the XY plane.)

Z-buffer   An area of memory holding the depth (Z) values of each surface as represented at each pixel location.

Zel   Occasionally used to refer to a unit of depth.

APPENDIX 2

FILE FORMATS

3D STUDIO

GEOMETRY FILE FORMATS

SHP 2D Shaper files of open or closed 2D polygons.
LFT Store the contents of the 3D Lofter.
3DS Contain all the data of the 3D Editor and Keyframer including mesh, animation and viewport configuration information. Meshes can be merged into any of the versions of the program with 3DS file format.
PRJ Contain all the information in all the modules and the most common way to hold work in progress.
DXF Data exchange format file used for moving files to and from CAD programs.
MLI Contain the material definitions from the Materials Editor.
FLI Low-resolution animation format. (320x200 or less)
FLC High-resolution animation format.

IMAGE FILE FROMATS

GIF Created by CompuServe, compact with fairly good image quality. Restricted to 8-bit colour depth.
CEL Old image file format for Animator Pro

Targa An old format, it was the first 24-bit format and was designed to work with Targa video boards. Fairly large files but can be compressed.

TIFF Common format often use to transfer images across programs and platforms. The most widely used graphics format on PC and MAC.

BMP File format developed by Microsoft for use with windows. 3D Studio can write 8 or 24-bit BMP files.

JPEG Joint Photographic Experts Group algorithm. A highly compressed still image file format.

MAX

GEOMETRY FILE FORMATS

MAX, SHP, 3DS, PRJ and DXF.
MAX 3D Studio MAX scene files containing all the information in the scene. (Like 3D Studio's PRJ)

IMAGE FILE FORMATS

Jpeg, Targa, TIFF, BMP, GIFF, FLC, AVI, EPS.
AVI Microsoft Audio/Video Interleave animation files.
EPS Encapsulated PostScript File format.

VIZ

GEOMETRY FILE FORMATS

DWG, DXF, MAX, SHP, LFT, 3DS, PRJ, VRML 2.0, STL
VRML Virtual Reality Modelling Language for use on the Internet
STL Stereo Lithography file.

IMAGE FILE FORMATS

JPEG, TARGA, BMP, EPS, TIFF, FLC/FLI, AVI

BIBLIOGRAPHY

SOME SUGGESTED READING

Computer graphics:
    Computer Graphics
        Vince (Design Council, 92)
    Computer Graphics - principles and practice
        Foley (Addison-Wesley, 90)

Particularly relevant to 3D:
    Fundamentals of Three-dimensional Computer Graphics
        Watt (Addison-Wesley, 89)
    Principles of Computer-aided Design
        Rooney (Pitman Publishing/Open University, 87)
    Computational Geometry for Design and Manufacture
        Faux (Ellis Horwood, 79)
    A Programmer's Geometry
        Bowyer (Butterworths, 83)
    Designing the future: The Computer Transformation of
        Reality   Baker *(Thames and Hudson, 93)*
    3D Studio for Beginners
        Lammers/Peterson(New Riders, 95)
    Inside 3D Studio 4
        Elliot/miller (New Riders, 95)
    3D Studio MAX Fundamentals
        Peterson (New Riders, 96)
    3D Studio MAX- Creating Hollywood-Style Special Effects.
        Bell (New Riders, 96)

On animation (particularly 3D):
    The Art and Science of Computer Animation
        Mealing (Intellect, 92)

For reference:
    Que's Computer Users' Dictionary
        Pfaffenberger (Que Corporation, 90)

Magazines:
    CAD/CAM *computer aided design & manufacture*
    XYZ *creative design & technology*
    Byte *general computing*
    Creative Technology creative digital design
    CGI   *(Computer Generated Imaging) creative digital design*

INDEX